THE YEAR IN SPACE

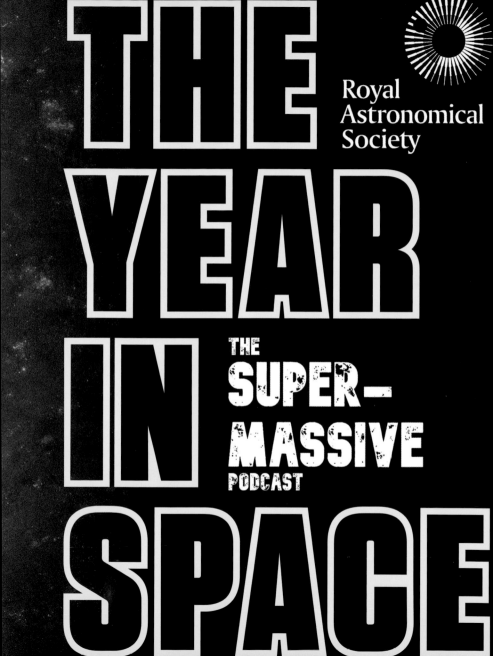

THE YEAR IN SPACE

Royal Astronomical Society

THE SUPER–MASSIVE PODCAST

First published in 2022 by
WILDFIRE
an imprint of HEADLINE PUBLISHING GROUP

1

Cataloguing in Publication Data is available from the British Library

Hardback ISBN 978 14722 9950 5

Designed by Amazing15

Diagrams on pages 130 to 145 © 2022 Greg Smye-Rumsby

Printed and bound in Italy by LEGO S.p.A

HEADLINE PUBLISHING GROUP

An Hachette UK Company
Carmelite House
50 Victoria Embankment London EC4Y 0DZ

www.headline.co.uk
www.hachette.co.uk

CONTENTS

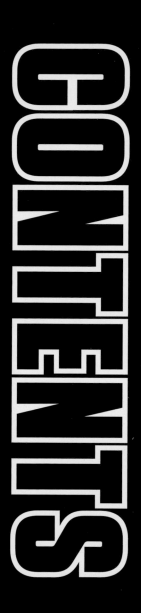

PREVIOUS PAGES: A stunning view of the spiral galaxy NGC 4571, taken by the Hubble Space Telescope in March 2022. *ESA/Hubble & NASA, J. Lee and the PHANGS-HST Team*

MEET THE TEAM

IZZIE CLARKE is a Webby Award-winning podcast producer, presenter and science journalist. Combining a Physics Masters with a background in entertainment radio, she has tried astronaut training, race-car driving and making fireworks in her mission to show people that science isn't boring. Izzie loves producing and presenting the Supermassive Podcast, alongside a variety of podcasts that cover everything from robotics to the environment (in her attempt to save the world ... one podcast at a time!).

DR BECKY SMETHURST is an award-winning astrophysicist and science communicator at the University of Oxford, specialising in how galaxies co-evolve with their supermassive black holes. She was recently awarded the Royal Astronomical Society's Research Fellowship for 2022. Her YouTube channel, 'Dr Becky', has over 500,000 subscribers who engage with her videos on weird objects in space, the history of science and monthly recaps of space news.

RICHARD HOLLINGHAM has reported for the BBC from Antarctica, the Libyan Desert and the top of a nuclear missile silo in Kazakhstan. A launch commentator for the European Space Agency (ESA) and mission control groupie, Richard has produced and presented numerous radio programmes and podcasts (some have even won awards). He writes for BBC Future and is proud to be the executive producer of the Supermassive Podcast.

DR ROBERT MASSEY is Deputy Executive Director of the Royal Astronomical Society. Robert started his career in astronomy in Manchester studying the Orion Nebula, worked as a teacher in Brighton, and spent eight years at the Royal Observatory Greenwich, where he developed an unbridled passion for bringing astronomy and space science to the wider public. These days he looks after the outward-facing work of the Royal Astronomical Society and enjoys the darker night skies of his family home in Sussex.

SUE NELSON is an award-winning science journalist, broadcaster and author of *Wally Funk's Race for Space*. Sue presents podcasts, makes short films for ESA and produces BBC radio documentaries – often with a focus on women in science and space. A former BBC science correspondent, she has ridden a replica Moon buggy with an Apollo astronaut, experienced the acceleration of space shuttle launch in a centrifuge, and flown thirty-one parabolas on a 'zero G' flight without throwing up.

SARAH WILD is a freelance science journalist. She studied physics, English literature and bioethics in an effort to make herself unemployable. It didn't work and now she writes about everything from cosmology to particle physics and all the steps in between. She's won awards, run national science desks and learned how to eat a sandwich while interviewing someone. Her work has appeared in *Nature*, *Science* and *Scientific American*, among other publications.

Aleta Harrison

Space is bloomin' big. Maybe even infinite. So how on Earth (and beyond) do you squeeze an entire year of space exploration into one book? It's a good question. It's certainly one that we asked ourselves when first approached to write the very thing that is now in your hands. But we've given it a go. (And don't worry, this book isn't somehow infinitely long ...)

If you don't know us, we're the space nerds behind the Supermassive Podcast from the Royal Astronomical Society. A podcast? Now on paper? Wild, we know. But if the past year is anything to go by, then that is by no means the most absurd thing that has happened.

For starters, a massive origami telescope unfurled in space. Then it started peering through the Universe to see the very first stars ever made (see page 8). We celebrate ten years of a rover roaming Mars in the search for life and investigate the plans to courier samples from the Red Planet back to Earth (page 58). But that's not all; we also look at the world's largest telescope operating out of two of the quietest corners of the planet, listening to the song of the Universe (page 42), and we meet the scientists planning the upcoming mission to Jupiter's icy moons (page 90).

See? We really have crammed a lot in.

If you've no idea who or what the Supermassive Podcast is, then let us explain ... We are the monthly podcast from the Royal Astronomical Society, produced by Boffin Media, and since our launch in 2020 we've become one of the most popular astronomy podcasts in the UK and US. Every episode, science journalist Izzie Clarke and astrophysicist Dr Becky Smethurst take listeners through the Universe with the latest research, history from the society's archives and astronomy you can do from home.

As for the Royal Astronomical Society, it has been at the heart of UK astronomy for over 200 years. The second largest astronomy society in the world, it has 4,000 members in seventy countries, consisting of professional and amateur astronomers, science writers, historians, teachers and people in the space industry. All are united by their love of space with many of them devoted to making it accessible to the public.

It's that same excitement for sharing the wonders of space that runs through the podcast and (hopefully) the pages of this book.

In fact, you might even notice a few similar features that have leaped out of the podcast and into these pages. Izzie is trying to shoehorn music and film references into all her space stories, Dr Becky is on hand to answer everyone's questions and Dr Robert Massey has all the stargazing and telescope tips covered.

Space is bloomin' big, but it's also endlessly fascinating. Let us show you why ...

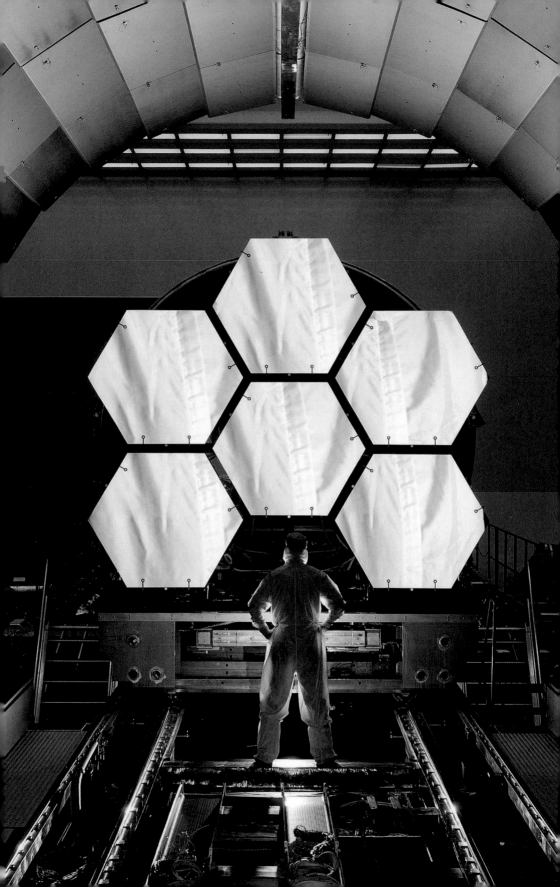

THE ORIGAMI TELESCOPE

The James Webb Space Telescope is revealing a new view of the Universe, says Izzie Clarke.

It was shaping up to be a Christmas Day like all others. Carefully wrapped presents were under the tree, the turkey was roasting in the oven, and there was the familiar sound of my godmother and Mum chatting away as they peeled carrots in the kitchen. But unlike Christmas(es) past, I have never sneaked away from The Big Day to watch a rocket launch.

Huddled on the end of the fold-out guest bed, phone in hand, I watched the small screen nervously as an Ariane 5 rocket blasted off from French Guiana. Folded inside was arguably the most precious cargo in recent astronomical history: the James Webb Space Telescope (JWST). As the largest, most powerful telescope ever launched, Webb will peer deep into the cosmos to see the formation of distant worlds. 'It was a relief,' says Mark McCaughrean, an interdisciplinary scientist at the European Space Agency (ESA). I'll be honest, it was not the response I was expecting. But it is understandable given that he has waited twenty-four years to see this observatory launch.

Talk of the telescope began in 1989, so McCaughrean sees himself as 'a latecomer' to the mission. 'We had seen the rocket the day before from close up and there was a real sense that you are part of it.' But when the launch day arrived there was what McCaughrean described as a 'tussle'.

"Then the sound arrived, and everybody went nuts. What I felt was just this sense of total disbelief."

'Our mission control is 12 kilometres away, and if you're the most senior person for science, you should be there.' But McCaughrean had other ideas. 'I said, "No, I'm going to be at the closest point I can possibly be," which is 5 kilometres away. There was a bit of a hullabaloo.'

McCaughrean stood with colleagues surrounded by Kourou's wild jungle, everyone happily chatting among themselves. He assures me that there was little sense of concern that it wouldn't work. And then the countdown began.

'The tension just went through the roof. And we've all heard so many Ariane 5 countdowns. You know exactly what each piece means. You know that zero doesn't mean lift-off. Zero means start the engines and then launch is six seconds later.' Looking out across the plain, a mixture of steam and smoke enveloped the rocket.

'I don't think I'd want to be any closer,' says McCaughrean. 'When the solid boosters lit up, suddenly there's this bright thing on the horizon.' The rocket moved quickly and a few cheers erupted as it climbed into the sky. 'Then the sound arrived, and everybody went nuts. What I felt was just this sense of total disbelief.'

Webb ascended quickly and was soon lost in the thick clouds. As a viewer, far removed from the inner workings of mission control, that's when my heart rate started to quicken. Was it on course? What if something had gone wrong? This is when McCaughrean's mind started racing too.

'We know what's coming next.' That 'next' stage meant unfolding the biggest space telescope ever built, cooling it down to cryo-genic temperatures, and peering into the early Universe. No biggie, right? As Robert Massey, Deputy Executive Director of the Royal Astronomical Society (RAS), put it, 'Ev-erything about it does sound a bit bonkers.'

But it has to be, to break barriers.

That seems to be the story with space telescopes. If you have ever seen any stunning space images, from swirling spiral galaxies to their oddball cousins – irregular galaxies – it is likely they've been taken by Webb's predecessor, the revolutionary Hubble Space Telescope (HST). But, at six times the size and half the weight, Webb is one hundred times more sensitive. The infrared images that it will take are going to transform our understanding of the Universe. It will penetrate clouds of gas and dust to capture the first stars to shine, hunt for habitable faraway planets, and look for that little thing called the origin of life.

The most 'bonkers' thing about this telescope is the design. It is big. Really big. Half the size of a Boeing 737 aircraft big. The sunshield is 22m (72ft) by 12m (39ft), similar to the size of a tennis court, and the large, hexagonal primary mirror towers above at 6.5m (21ft). And the only way to get this complex structure into space was to origami-fold it into a rocket and have it unfold on its way to its destination.

PAGE 8: JWST's primary mirror segments are prepped for testing. *NASA/ MSFC/David Higginbotham*

OPPOSITE: Webb safely stowed away on an Ariane 5, poised for launch the following day. *ESA/CNES/ Arianespace*

LEFT: JWST launch, 25 December 2021. *NASA/Bill Ingalls*

TESTING, TESTING

There is no denying that it has taken a lot of work to get to this point. As a huge international collaboration between NASA, ESA and the Canadian Space Agency, the creation of Webb has involved more than 300 universities, organisations and companies across twenty-nine US states and fourteen countries.

Initial hopes were that Webb would take flight by 2010, some two decades after planning began, and would cost $1 billion. Having launched in 2021 and at a price of $10 billion, it's safe to say that everyone's patience and purse strings were put to the test. There were engineering faults, political hesitancy, and seemingly endless delays. And delays mean money. The US Government Accountability Office (GAO) was keeping a keen eye on progress and its annual report makes for interesting reading.

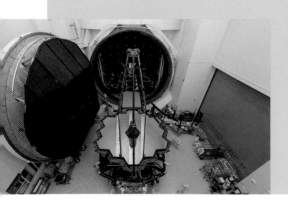

"There is no denying that it has taken a lot to get to this point."

In May 2017, for example, there was an investigation into leaks within the spacecraft's propulsion valves. Turns out a technician had cleaned them with the wrong solution. To make matters worse, someone else then applied too much voltage to the refurbished modules, resulting in a two-month delay. Another exercise in October 2017 was to unfold, or deploy, the vast sunshield. The sunshield peeled back to reveal several tears. Further investigations concluded these were due to 'workmanship error'. Two more months of delays.

Then came April 2018. JWST was a few months ahead of its (optimistic) launch date and went through 'environmental tests'. These are pretty standard to ensure the thing you have spent decades making can survive the ride into space. They did not go well. Screws and washers – physical hardware to keep the sunshield in place – had broken free, scattering around the test chamber. At that point, the launch was delayed for another three years. You can understand McCaughrean's sense of relief when launch day finally arrived in 2021 – although, true to form, it was six days later than planned.

OPPOSITE TOP:
The primary mirror segments are gold-coated. *NASA/ MSFC/David Higginbotham*

OPPOSITE BOTTOM:
JWST ready for testing in a space simulation chamber at NASA's Johnson Space Center. *NASA/Desiree Stover*

BELOW: Cleanroom workers watch the mirrors being rotated so that the science instruments can be installed behind. *NASA/Chris Gunn*

FAILURE NOT AN OPTION

All things considered, Webb had a smooth launch. So smooth that McCaughrean even described it as 'completely perfect'. When planning a mission as bold as this, scientists allow for a window of possible error that may tweak a trajectory but still give the result you want. Webb, however, followed a path that simulations would have described as flawless. So far, so good.

Next came the deployment of the sunshield and unfolding the telescope. For those scientists who had spent decades on this mission, this is where the nail-biting began. Think about it. You have a 6.5m (21ft) dish that is made up of eighteen individual mirrors. Each of those has pins to release mechanisms – the primary mirror itself has 178 release mechanisms alone, plus additional latches to keep things in place. In fact, JWST had 344 'single points of failure' – consider them breaking points – when it left Earth. All it needed was for one of those to kick off and it was mission over.

The first step to unfolding the biggest telescope ever made began thirty-three minutes after launch with the release of Webb's solar array, to enable it to generate its own power. Fortunately, the manoeuvre went smoothly. A day later, the communication antenna sprang into action. After these two automatic processes, everything else was controlled by the team on the ground. All eyes were on Webb.

Two days after launch, the observatory still looked like a bud waiting to bloom. The primary dish was curled up, enveloped by two structures carrying the sunshield. But by day three the sunshield started to peel away, beginning its long-awaited deployment phase. Two long, rectangular planks – officially called the Unitized Pallet Structures (UPS) – carrying the folded-up sunshield membranes, unfurled at the front and back. A day later, the now-exposed telescope rose by 1.22m (4ft) to distance the telescope from the spacecraft – firstly, to create room for the sunshield to stretch out when the time came; and, secondly, so that the telescope could begin its cooling process.

Six days after launch and JWST's sunshield now started to resemble the famous kite structure we had been promised. With every completed step, the likelihood of having a new space telescope was increasing. A few days into the New Year and the two-day activity of tensioning the sunshield began. Three of the five layers were pulled into place, and by the following day all five layers were fully tensioned. One enormous parasol had been deployed. After decades of delays and increasing costs, this was the first proof that Webb was worth it.

If you have lost track of the days, we are now at day eleven. But that period after Christmas is always hazy. Now it's the telescope's time to shine. From an upright position, secured by long metal arms (aka booms), the secondary mirror swings in front of the twelve central mirrors. This smaller mirror is responsible for reflecting everything the mammoth main dish collects to beam it to the instruments on board. Finally, the wings of the primary mirror unfurl. Right goes first and it's a success. Left wing follows ... and exactly two weeks after launch, the observatory is in place.

"Webb assumes the position of the most impressive space telescope ever made."

Webb assumes the position of the most impressive space telescope ever made.

If space wasn't a vacuum, it probably could have heard the cheers from ground control.

Webb arrived at its destination on 24 January 2022. Lagrange Point 2 (L2) is located 1.5 million kilometres (932,000 miles) from our blue dot. It is a fortuitous region, in the exact opposite direction from the Sun, where the gravity from the Earth and our star balances the orbital motion of a satellite. Parking Webb here gives it a fixed position relative to the two bodies.

Being quite literally a million miles from Earth is also an ideal spot to radiate heat out into deep space. Hunting for those earliest galaxies is no easy job. They are old. And that means their infrared (heat) signals are faint. The last thing Webb needs is our local fireball, with its light and heat, throwing it off the trail. The fact that the telescope is plunged into twenty-four-hour darkness at temperatures just above absolute zero gives it the best chance possible.

So how does the James Webb Space Telescope actually work?

DUSTY VEIL

'We always start off with talking about the Hubble Space Telescope,' Keith Parrish, Observatory Commissioning Manager for the James Webb Space Telescope, told the Supermassive Podcast. I know you are here to read about Webb but bear with.

'Hubble is just fantastic,' continued Parrish. 'One of the most ground-breaking things that Hubble has taught us is that the amount of galaxies in our Universe is mind-boggling.'

Launched on 24 April 1990, HST delivers 140 gigabytes of science data back to Earth every week and has transformed our view and knowledge of the Universe. 'In every direction that you would look, Hubble's actually able to spy some of these early galaxies,' said Parrish. 'They just look like little red dots. Nobody knows how they're constructed.'

ABOVE: Hubble during a servicing mission in 2000. The astronauts give a sense of scale. *NASA/ESA*

LEFT: The giant nebula NGC 2014 and its neighbour, as captured by Hubble in 2020. *NASA/ESA/STScI*

Hubble has shown us so much more than these captivating little red dots. It has revealed the size and age of the Universe, the birth and death of stars, and the formation of galaxies and storms on Saturn and Jupiter. It has been used to investigate dark matter and probe the atmospheres of alien worlds. And alongside these leaps of knowledge, it has left a huge cultural legacy.

Hubble itself isn't exactly small. NASA describes it as being the size of a large school bus, and having the weight of two adult elephants (if you are inclined to measure things in elephants). For the stargazing fans, it's an epic Cassegrain telescope design. For

everyone else, it's tube-like in shape.

What sets Hubble apart from telescopes here on Earth is that it sits above our atmosphere, the very definition of a 'space' telescope. At just 547km (340 miles) above Earth, Hubble can observe objects that would be impossible to see from the ground. Although the telescope does have some infrared capabilities, it mostly studies the Universe at optical and ultraviolet wavelengths, connecting us to its infinite vastness. But JWST will send us further.

Let me take you back to those 'little red dots' that Keith described. Hubble can capture some of those infrared signals,

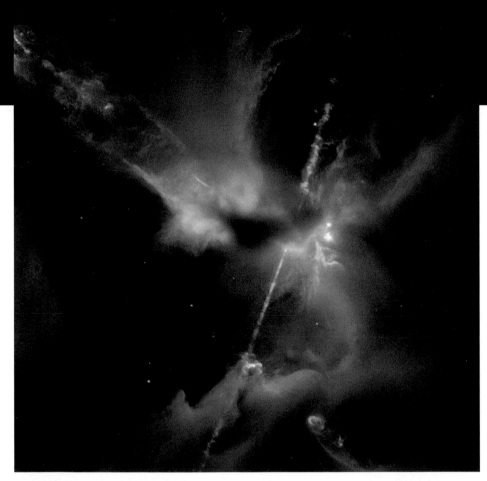

ABOVE: Jets of energised gas ejected from a young star. *ESA/Hubble & NASA, D. Padgett (GSFC), T. Megeath (University of Toledo), and B. Reipurth (University of Hawaii)*
OPPOSITE: The star cluster Pismis 24. *NASA, ESA and Jesús Maíz Apellániz (Instituto de Astrofísica de Andalucía, Spain)*

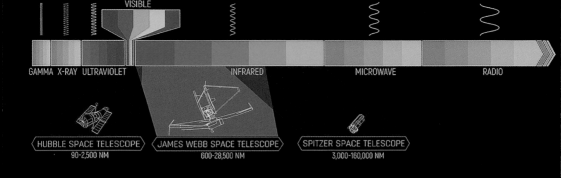

"These hidden realms are where planets and stars form."

as long as the light they emit sits somewhere between 0.8 and 2.5 microns on the spectrum of light. But what it cannot tell us is what is hiding behind the clouds of gas and dust that may surround them. These hidden realms are where planets and stars form. And they absorb visible light. Hubble cannot penetrate them. Webb, an infrared telescope, will carry on Hubble's legacy and cut through that dusty veil.

Parrish was understandably very excited about the new telescope's capabilities. 'We're going from that very first thing: how did stars and galaxies come about?'

Webb can, hopefully, answer that question by studying their light. The older the star, the further away it is, and, therefore, the more 'stretched' or 'redshifted' its light becomes. Essentially, their signal is pushed from UV and optical wavelengths into the near-infrared as it travels away from us.

'Then we can go down another level. How do planets form?' said Parrish. 'I always say George Lucas had it right in *Star Wars*. He knew that there were lots of planets out there. Almost every star you look at in the sky probably has a planet around it, and that's really, really cool.'

We already know that Tatooine-like systems exist, with planets circling two stars. But let's not forget, at the time of Hubble's launch, we did not even know that there were other planets beyond our Solar System.

Parrish's list of potential JWST benefits continued to grow. 'And more importantly, how do planets with all the conditions that we would expect to support life form? To think that we have built a telescope that's a hundred times more sensitive than Hubble. I mean, it's hard for me to get my head around and I work on it every day.'

KITCHEN FOIL

To explore these new horizons, Webb needs its sunshield, telescope and instruments to work in unison, operating at 'beyond-frostbite' temperatures to transport us to the depths of the unexplored Universe. It's dark, cold and big out there.

Aerospace company Northrop Grumman has put materials science to the test to start that journey.

Take the iconic sunshield, for example. Its tennis-court-sized parasol is made from a wonder lightweight material called Kapton, with a coating of aluminium for that kitchen-foil glimmer. Across five layers, temperatures start at 383K (110°C/230°F) on the Sun-facing side and drop down to a chilly 36K (−237°C/−394°F) on the cold side.

The thing I struggle to wrap my head around is just how drastically the temperature is reduced across five thin layers. As sunlight hits the first layer, a lot of it is radiated back into space with the added help of a doped-silicon coating. Some heat will get through but then the second layer blocks and redirects that away too. With each layer, heat continues to be dissipated. Layer one is just 0.05mm thick and the rest are 0.025mm. And if you are thinking about the coatings, well, they can be measured in nanometres.

BIG DISH

Then we have the primary mirror. The bigger the mirror, the more light it can collect, and the more detail it can see. The engineers designing Webb worked out that a diameter of 6.5m (21ft) would do the trick. They then just had to make it, jam it into a rocket and get it into space.

Cue the second wonder material: beryllium. A third lighter than aluminium but six times stronger than steel, and with the ability to keep its shape when plunged into cryogenic temperatures, this beat other options hands down.

JWST's sunshield fully unfolded during testing at the Northrop Grumman facility, California. *NASA/Chris Gunn*

This audacious primary mirror consists of eighteen smaller hexagonal segments, each 1.32m (4.3ft) in diameter, which pack together like honeycomb. Going from raw material to shiny completed segments has taken fourteen stops to eleven places across America. Each mirror started its life as a hefty 250kg (550lb) hexagonal cast, or 'blank'. But there is no way they could launch eighteen of those; the fuel cost would be ridiculous. The manufacturers were able to remove 92 per cent of that material by carefully cutting out intricate triangular patterns from the back, bringing it down to just 21kg (46lb). Next came the polishing and processing, to ensure each mirror was as close to perfectly smooth as possible – reaching accuracies of less than one millionth of an inch.

The final touch to the design was a microscopically thin layer of gold. Why? To make it shiny. Or, in engineering terms, 'to ensure the highest infrared light reflectivity to Webb's instruments'. Silver sits at 95 per cent reflectivity, aluminium is about 85 per cent, whereas gold hits the jackpot at 99 per cent. It is also extremely unreactive. We cannot have the world's largest and most powerful infrared observatory tarnishing in space, can we?

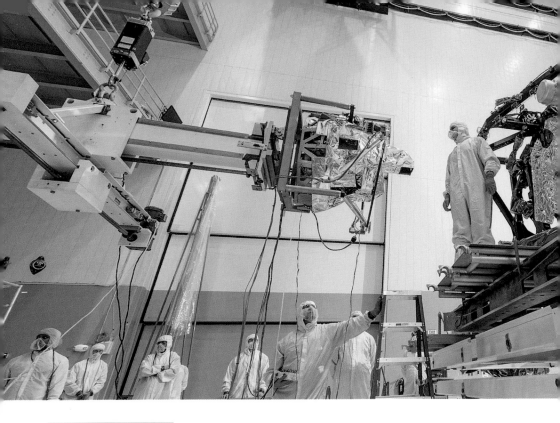

STAR HUNTER

Directly behind Webb's primary dish is the Integrated Science Instrument Module (ISIM) which houses four key instruments. The telescope collects the light, which is beamed to all of them at the same time by the primary mirror, ready to pick apart the mysteries encoded within.

First up, we have JWST's primary imager, the Near-Infrared Camera (NIRCam), built by the University of Arizona. Consider it the galaxy and star hunter. Operating at wavelengths between 0.6 and 5 microns, it can detect light from the formation of the earliest stars and galaxies, as well as young stars in the Milky Way and Kuiper belt objects. To do so, it simultaneously scans two adjacent fields of view and observes them in both shorter and longer infrared wavelengths. It is also equipped with something called a 'coronagraph'. This blocks starlight, allowing us to examine smaller, fainter objects nearby, like planets. It is not dissimilar to holding your hand to your head on a sunny day,

allowing you to focus your view – hopefully upon the ice-cream van that's approaching in the distance.

Then there's the Near-Infrared Spectrograph, or NIRSpec. The chemist. NIRSpec splits the light from far-off stars into a spectrum to reveal the object's physical properties. Just like a unique fingerprint, an object's light can leave a pattern that is a signature of individual chemical elements, revealing what they're made of, as well as temperature and mass.

NIRSpec can observe 100 objects simultaneously, thanks to 250,000 microshutters. These tiny windows with shutters, the width of a human hair, open and close when a magnetic field is applied and can be controlled individually. Never have scientists been able to analyse so many objects at precisely the same time.

After that, there is the catchiest name yet ... the Fine Guidance Sensor/Near-Infrared Imager and Slitless Spectrograph (FGS/

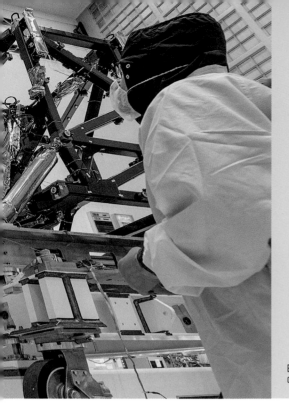

Engineers installing MIRI into the Integrated Science Instrument Module, in the cleanroom at NASA's Goddard Space Flight Center. *NASA/Chris Gunn*

NIRISS), provided by the Canadian Space Agency. The guider. FGS will help point the telescope and its NIRISS component is responsible for investigating the first light detection, exoplanets and their characteristics, as well as studying their spectroscopy as they move past their star, otherwise known as a transit. Finally, there is the Mid-Infrared Instrument (MIRI), made by a British team. The double whammy.

'It is both a camera and a spectrometer. That makes it a little bit special,' Gillian Wright, MIRI's principal investigator, told us. 'It's really important because you break the light down into its constituent colours and every atom in the Universe has a unique signature. In the mid-infrared where MIRI looks, it's very sensitive to seeing interesting types of molecules like methane and water.'

While MIRI, too, will be working with the other instruments, looking at the first stars and exploring galaxies, there's something that makes it unique.

'We're studying the structure in those disks [clouds of gas and dust] and trying to understand which disks would have the potential to make what kinds of planets,' said Wright. 'And MIRI is very good at that. Because of its long wavelengths, we can peer much further into these dusty disks than you can with the other instruments.'

These long wavelengths are 5–28 microns, which is great for cutting through cold, dense and dusty regions that had previously been closed off to prying telescopes. Plus, if we want any hope of finding the oldest galaxies, it will be MIRI that will spot their distant, stretched (redshifted) light. Putting it into perspective, mid-infrared light is given off by objects at room temperature, so these signals are going to be faint. To ensure MIRI snaps them up, it comes with its own fridge, cooling the instrument to just 7K (–266°C/–447°F), keeping it clear of any background light.

FIRST LIGHT

People waited in anticipation for the first images from Webb. I would not be surprised if sweepstakes were running through the astronomical community. We had our own (hypothetical) money on Webb's take on the Pillars of Creation, three towers of gas and dust in the Eagle Nebula made famous by Hubble in 1995. But before those first images appeared, NASA kept the telescope's fans in the loop every step of the way.

'They didn't really intend to,' says Mc-Caughrean. I wondered if that was because NASA faced so much ridicule after Hubble launched. Initially, the $1.5 billion Hubble Space Telescope was a disaster. Most serious

among a catalogue of flaws – including communications, component and electronic failures – was a fault with the primary mirror. Hubble became the butt of jokes across the world.

Mark mentioned there were concerns regarding Webb's earliest data. 'If we show them not very good images, they'll say it's broken.' I imagine NASA initially wanted to keep their cards close to their chest.

But, internally, there was pushback. 'Take them on the journey. Take them through the process. Eighteen mirrors. They're not super-sharp. Now you align them and just have to tell the story.' And thank goodness they did. Every calibration, every 'rough' image, was celebrated.

One of the earliest alignment phases for Webb was from NIRCam. It produced an image of eighteen smudges; a single point interpreted by each individual mirror. By making tiny adjustments, as well as with the secondary mirror, those eighteen smudges transformed into eighteen clear points. With the segment alignment completed, the actuators – a series of small motors controlling the mirrors – moved each segment to overlap their light and work as one. The engineering to do this did not exist before Webb. We are talking seriously small movements here. Engineers moved the mirrors by 1/10,000th the thickness of a human hair. It's still bonkers.

By 16 March 2022, we had our first test image. 'Following the completion of critical mirror alignment steps,' stated NASA's press release, '[the] James Webb Space Telescope team expects that Webb's optical performance will be able to meet or exceed the science goals the observatory was built to achieve.' This was the confirmation that Webb was going to deliver.

Suddenly, this unspecific star that just happened to be the right brightness for calibration became one of my favourite images of space. And get used to seeing those eight shining beams emerging from a point, this is Webb's signature look. The six large ones are the diffraction pattern from its hexagonal shape, with two smaller lines from Webb's secondary mirror.

That favourite image, however, was blown out of the water in July 2022. From the deepest view of the Universe to cosmic mountains and valleys full of twinkling stars, NASA released the first five images from JWST and our view – and understanding – of the Universe was changed forever. The James Webb Space Telescope had given us a view into the unknown and it was everything I had hoped for.

It was also just as McCaughrean had predicted:

'We know the things that it's designed to do, and we know they haven't been touched for two decades because you just can't get close to them without Webb,' he says. 'It's all still there waiting but what else is there to find? Could it possibly be that there's something completely different that we haven't even thought of? I can't tell you what they'll be, but it'll be amazing when they come up.

'And with a machine this powerful, they will.'

ABOVE: The same star, as seen by the Spitzer Space Telescope (launched in 2003) and JWST. *Left: G. Brammer (University of Copenhagen); Right: NASA/STScI*
OPPOSITE: A new shot of the iconic Pillars of Creation image taken by Hubble in 2015. *NASA, ESA/Hubble and the Hubble Heritage Team (STScI/AURA)*

REVEALING THE WONDERS OF THE EARLY UNIVERSE

After the excitement of launch and successful deployment, the first pictures from the James Webb Space Telescope were always going to be special. Only a select few within the space community had any idea what these images would be. It was such a closely guarded secret, the rest of us could only speculate.

When the day came, the pictures did not disappoint. US President Joe Biden unveiled the first image during a White House briefing on 11 July 2022. The next day, in a global broadcast, NASA and ESA released a further four images (including a graph showing the atmospheric composition of the exoplanet WASP-96b, not pictured here), giving a perspective that NASA administrator Bill Nelson described as 'a view the world has never seen before'.

Over the coming months, we can expect plenty more revelations from Webb, but these first images will always remain significant – proof that if we humans put our minds to something, we can do incredible things.

ABOVE: Webb's view of Stephan's Quintet, a compact group of galaxies located in the constellation Pegasus, will give scientists new insights into supermassive black holes and how galaxies interact. The image contains over 150 million pixels and is constructed from almost 1,000 separate files. *NASA, ESA, CSA, and STScI*

LEFT: This seemingly three-dimensional image of the Carina Nebula – nicknamed the Cosmic Cliffs – looks like a landscape of glittering mountains and valleys. In reality, it is the edge of a giant star-forming region, buffeted by the blistering ultraviolet radiation and stellar winds emanating from the young stars in the centre of the nebula. The tallest peaks in this image are about 7 light-years high. *NASA, ESA, CSA, and STScI*

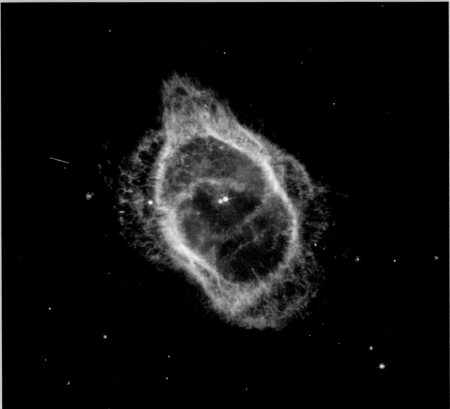

OPPOSITE: This planetary nebula – an expanding cloud of gas that surrounds a dying star – is approximately 2,500 light-years away. Known as the Southern Ring Nebula, it is revealed here in all its beauty by both Webb's Near-Infrared Camera (top) and its Mid-Infrared Instrument (bottom). *NASA, ESA, CSA, and STScI*

ABOVE: Galaxy cluster SMACS 0723 as it appeared 4.6 billion years ago. The image is teeming with thousands of galaxies – including the smallest, faintest objects ever observed , some of which are seen as they were just a few hundred million years after the Big Bang. The image was captured over 12.5 hours by Webb's Near-Infrared Camera. *NASA, ESA, CSA, and STScI*

THE LAST MAN ON THE MOON

It's fifty years since Gene Cernan became the last man to walk on the Moon. Richard Hollingham shares a unique conversation with the late NASA astronaut. It covers a career that includes a 'spacewalk from hell', a near-death experience at the Moon and the final Apollo mission to the lunar surface.

ABOVE: Gene Cernan in 1971. *NASA* OPPOSITE: Apollo 17 EVA. *NASA*

Shortly after 05:30 GMT on 14 December 1972, the commander of Apollo 17, Eugene Cernan, prepared to step off the lunar surface.

During this final Apollo mission to the Moon, Cernan and his companion, geologist Dr Harrison Schmitt, had set records for the longest time spent on the Moon (seventy-five hours in total, twenty-two of those outside), the distance travelled in the lunar rover (35km/22 miles), and the amount of rock they were bringing back to Earth (111kg/244lb). By every metric the expedition was a success.

With the world watching, Cernan knew that his final words spoken on the lunar

> "As we leave the Moon and Taurus-Littrow, we leave as we came, and, God willing, we shall return, with peace and hope for all mankind."

surface – just like Neil Armstrong's first – would go down in history. But he had no idea what he was going to say.

'I'm more of an ad-lib guy,' he told me. 'I tend to put the words together as they come to me. I, honest to God, did not know what I was going to say until we left.'

It is perhaps why the final words spoken on the Moon sound so heartfelt.

'America's challenge of today has forged man's destiny of tomorrow,' said Cernan. 'As we leave the Moon and Taurus-Littrow, we leave as we came, and, God willing, we shall return, with peace and hope for all mankind.'

What saddened Cernan – even angered him at times – was that he never got to see

NASA's plans take shape to return humans to the Moon (see page 118). For the rest of his life, and to this day, Cernan remains the last man on the Moon.

The interview that follows was recorded at an event called Spacefest (which very much does what it says on the tin) in Tucson, Arizona, in the summer of 2016. It was one of Cernan's last interviews. He died, aged eighty-two, a few months later.

Both suffering from jet lag, we met in his hotel room at seven in the morning and chatted for almost two hours. I had met Cernan a couple of times before but on this occasion he was more contemplative, happy to talk about the highs and lows of his extraordinary career as an astronaut.

Growing up in a small town near Chicago, the son of Czech and Slovak immigrants, Cernan graduated from high school before studying electrical engineering at university. But he had always wanted to fly. Realising his ambition, he qualified as a captain in the US Navy and flew fighter jets off aircraft carriers; he also earned a master's degree in aeronautical engineering. With some 200 landings on moving ships behind him, he was selected as an astronaut in 1963.

Over three spaceflights, spanning six years, Cernan became the third person ever to walk in space, was one of only three astronauts to journey to the Moon twice,[1] and was the only astronaut to twice fly a lunar lander.

As a member of the crew of Apollo 10 – along with Tom Stafford and John Young – he also holds the record for the fastest re-entry to the Earth's atmosphere and, therefore, the record for the fastest speed ever attained by humans. In May 1969, they returned home at 39,937km/h (24,816mph).

He deserves to be remembered.[2]

----

[1] The others were Jim Lovell (Apollo 8 and 13) and John Young (Apollo 10 and 16).

[2] I have made only minor grammatical and punctuation edits to Cernan's words and have sometimes abridged his answers. For clarity, I have added additional information to some of my questions.

OPPOSITE TOP: The crew of Gemini 9, Thomas Stafford (left) and Eugene Cernan, January 1966. *NASA*

LEFT: Cernan practising with his 'Buck Rogers backpack' ahead of the Gemini 9 mission. *NASA*

"We didn't know beans about going to the Moon. We had a whole reservoir of things we had to learn."

RICHARD: I'd like to start by talking about your first mission in 1966, Gemini 9. Gemini was a two-man spacecraft designed to prove some of the technologies needed to go to the Moon. Can you give us a sense of how important the Gemini programme was?

GENE: We didn't know beans about going to the Moon. We had a whole reservoir of things we had to learn. That's when Gemini came in. Gemini was the bridge, and a very important bridge, that we had to cross successfully to get to Apollo. We needed to learn how to rendezvous, we needed to learn how to spend long durations in space. We needed to learn how to get out of spacecraft and confront the elements of not just zero gravity, but the vacuum of space. All the things that we were going to have to confront and learn how to do to walk on the Moon.

Did we have some failures in Gemini? Yes. Do you learn from your failures? Hopefully you do.

RICHARD: From the off, your Gemini 9 mission had failures – you and Commander Tom Stafford were due to practise rendezvous and docking manoeuvres with an unmanned spacecraft, the Agena target vehicle, but that plunged into the sea shortly after launch. A replacement docking target then reached orbit but failed to deploy. When you finally launched and reached orbit, you found that you couldn't dock.

GENE: Tom Stafford nicknamed it the angry alligator. The jaws were open, it was held on by a band, and there was a lot of discussion over whether I should get out of the spacecraft, spacewalk and try to cut that band off. A lot of people said it would be damn-near suicide because as soon as you cut the band that thing's going to snap and where's it going to go? It's probably going to hit me somewhere. The decision was made not to do it, which was probably a good decision. And so, we just rendez-voused on that vehicle, which we were going to have to do in Apollo. And then the last day [of the Gemini 9 mission], it was my spacewalk, which was a whole other interesting deal.

RICHARD: This would be only the second American spacewalk, or extravehicular activity (EVA), and it was hugely ambitious. You had to leave the Gemini capsule, spacewalk around to the rear and attach yourself to a device called the astronaut manoeuvring unit. You've called it 'the spacewalk from hell'.

GENE: You know, as smart as we were, as engineers, we're in zero gravity and I had no tethers to hold me down. I was standing on a handlebar in zero gravity. You don't stand in zero gravity. So, I had to basically assemble, as I call it, this Buck Rogers backpack. I was going to fly around out there on a 38m (125ft) tether. I had to turn the fuel valves on but every time I turned a valve in zero gravity, it turned me. So here I am, turning the valve and I'm floating back out there and trying to hold on with one hand. I had to put one foot under the handlebar and one on top just to hold myself on. It was very difficult. I ended up overpowering the environmental control system; my helmet fogged up so I couldn't see. I also ripped a bit of my suit and I could feel the burning Sun come through the thermal layers. It was an interesting spacewalk.

RICHARD: And the astronaut manoeuvring unit was also potentially dangerous.

GENE: I wore chrome steel woven pants. Should have caught that one right away. Why the hell? Well, because this particular rocket pack had hydrogen peroxide rockets in it, which I was going to use to manoeuvre myself around. You fire a rocket, boom, there goes the fire. And someone said, we can't let him catch on fire. Maybe we'll make steel pants. You know, there's so many things we didn't catch, and I probably should have caught most of them.

RICHARD: You were having so much difficulty that, in the end, Tom Stafford and mission control called off the spacewalk

and ordered you back inside. But even squeezing back into the spacecraft was a challenge.

GENE: My heart rate was going 170 beats a minute back there. On the ground the doctors were going bonkers – they'd got an astronaut going around the world with 170 beats a minute. They knew I was in trouble. I've always said I didn't go to the Moon not to come home, and I didn't get out of that Gemini spacecraft not to get back in – as hard as I knew it would be.

I was bent like a pretzel, every bone of my body bent one way or another. And it hurt. It hurt. Tom finally pressurised the spacecraft then my suit went down, and I could take a breath. He said I was so solid red, I looked like a turnip. When I got my helmet off, he took the water nozzle and he just squirted me with it.

There're so many things we missed on Gemini 9 that we could have thought of ahead of time. The other side of that coin is that the more successful you are, the more complacent you get. I always looked at my EVA as a failure, I let the guys down. In retrospect, over time, we probably learned more because of the problems I had than we would have if everything had gone successfully.

RICHARD: What would have happened if you couldn't get back into the spacecraft ... was there a plan?

GENE: Yeah, there was. But we didn't talk about it much. Tom would have had no choice. He would have had to come home without me. He would have disconnected the umbilical and I'd still be floating up there somewhere. It sounds terrible, but you know, supposing we couldn't get guys off the surface of the Moon? There was no rescue ship. We had to accept the fact that there was some risk.

RICHARD: Your second mission was Apollo 10 in May 1969, billed by the media

as the 'dress rehearsal' for the Apollo 11 landing later in the year. Three of you flew to the Moon – John Young remained in the command module as you and Tom Stafford took the lunar lander to within just 14km (8.7 miles) of the lunar surface. But it was during your return to the command module that things went wrong.

GENE: The mission was going great. But as many times as we practised what we were doing, human error can always come in. I flipped the switch over here on the left. Tom knew which switch had to be changed [for the manoeuvre] and he takes it and flips it over the other direction. And the spacecraft spun out of control. I saw the lunar horizon go by in different directions eight times in fifteen seconds. Is that scary? Yeah, if you get time to be scared, but I didn't have time to be scared.

Tom finally was able to take over manually. And, God bless his soul, he literally saved my life because they told us afterwards had we gone around once or twice more, two things would have happened: we would have gone into what they called gimbal lock, where all our navigation equipment froze up and you're really in trouble then. Or we would have taken enough

energy out of our orbit that we would have slowly drifted down to the surface. So, you've got to be careful that you don't get too overconfident.

RICHARD: Let's talk then about Apollo 17 – the culmination of the Apollo programme. You and Harrison Schmitt drove some 35km (22 miles) across the lunar surface, you got so much science done and you left the final footstep. You must be proud?

GENE: I was an underdog, the guy that didn't go to test pilot school. I needed that flight to prove that I was good enough. I was willing to literally turn down an opportunity to walk on the Moon [during an earlier selection] for maybe a chance to command Apollo 17. And that's what I felt when I stepped on the Moon. It wasn't the steps that counted. It was the fact that I proved to myself that I was good enough to get that far.

Harrison Schmitt standing next to a lunar boulder during the Apollo 17 mission, 13 December 1972. *NASA/Eugene Cernan*

"It wasn't the steps that counted. It was the fact that I proved to myself that I was good enough to get that far."

23

"We had to come back with some answers to questions we didn't even know that we were smart enough to ask."

OPPOSITE: Cernan drove some 35km (22 miles) in the Apollo 17 lunar rover. *NASA*

ABOVE: The astronauts unfolded the buggy from the side of the lunar lander. *NASA/Harrison Schmitt*

I was on a high the whole time. Everything seemed to work pretty well because every flight we had built upon the previous flight, built upon its successes and built upon the failures. And when we got to Apollo 17, I think we were really prepared. And from a scientific point of view, I took a scientist and they say this is the best science flight we've ever had. Just going to the Moon and picking up a rock was not going to be good enough. We had to accomplish and come back with some answers to questions we didn't even know that we were smart enough to ask.

RICHARD: You're a firm advocate for a return to the Moon. What can people – astronauts – do on the Moon that robots can't?

GENE: The robot is only as smart as we make it before we send it. This brain right here is the most complex computer in the world. It's human courage, human culture,

human ability to think, to understand what discovery is. We're curious, you can't make a robot curious. And curiosity is the essence of human existence. What's on the other side of that hill? What's across the river? Now, a robot may not care, and it may not ask the question, but you do. When have we had a ticker-tape parade for a robot?

RICHARD: You've spent most of the latter part of your career taking part in education programmes, talking about your career, and you even starred in a

documentary film about you, *Last Man on the Moon,* **which was met with glowing reviews around the world.**

GENE: Young kids have to see that these guys who went to the Moon didn't come out of the sky in silver capes. You know, they put on their pants one leg at a time and had a family, they had kids. They made mistakes. There was a message I felt almost responsible to leave with these kids. If I can do it, why can't they? And if I can get them to try, then maybe they'll realise how good they can be.

"It's human courage, human culture, human ability to think, to understand what discovery is. We're curious, you can't make a robot curious. And curiosity is the essence of human existence."

OPPOSITE: The crew of Apollo 17 captured the famous Blue Marble picture of the Earth, depicting Africa at its centre. *NASA* **BELOW:** Apollo 17 is still producing useful science. In 2019, NASA scientists opened an untouched rock and soil sample from the mission to analyse it with technologies not available in 1972. *NASA/James Blair*

LISTENING TO THE SONG OF THE UNIVERSE

From the Moon to MeerKATs, Sarah Wild explores the challenge of building the world's largest radio telescope.

f you really want to get away from it all, then the far side of the Moon is hard to beat. Standing on the edge of a rugged crater, you would be almost completely shielded from the cacophony generated by humanity on Earth. It's a perfect location to listen to the radio waves whispered by celestial objects.

Nearly all things you can see in the cosmos – and many you can't – emit radio waves. Some, like exploding stars, scream their presence into the void, while others, like the first galaxies that formed after the Big Bang, only murmur their secrets to those searching for them. Radio telescopes detect these weak signals and piece them together to listen to the song of the Universe.

But even the loudest astronomical object is only a slight mumble when pitted

against a mobile phone or a nearby car, which is why the far side of the Moon is so ideal. Unfortunately, shipping thousands of tonnes of radio telescope to the Moon is not particularly practical, to say nothing of the cost and carbon footprint.

So, when the team behind the Square Kilometre Array (SKA) started hunting for a site to host the world's largest radio tele-scope, they looked instead to some of the quietest corners of Earth, places where hu-mans were few and economic activity lim-ited. They found them in the dry plains of South Africa and Australia. Despite being more than 10,000km (6,200 miles) apart, the two locations will each host a piece of the world's largest radio telescope.

When complete, the SKA will comprise about 2,000 giant dishes on the African continent and up to a million Christmas-tree-like antennas in Australia, all working together to make up – you guessed it – 1km^2 (0.4 square miles) of collecting area. In these remote places in the southern

hemisphere, with only panoramic skies for company, the SKA will be able to answer some of astronomy's major open, and most fundamental, questions: Are we alone? How are galaxies formed? And what happened just after the Big Bang?

After almost three decades of planning, the SKA Observatory has started awarding tenders for construction of the €1.9 billion first phase of the project. In 2019, seven founding nations – Australia, China, Italy, the Netherlands, Portugal, South Africa

and the UK – established the observatory and, recently, Switzerland joined as the eighth member. Numerous other countries, such as Germany and France, are participating through their scientific organisations.

Meanwhile, providing a tantalising preview of SKA's potential, in South Africa's Karoo hinterland, sixty-four dishes three storeys high – collectively known as the MeerKAT telescope – are already listening to the secret sounds of the Universe.

COMPARE THE MEERKAT

The name MeerKAT sounds like a clever marketing ploy, but it came about more by accident than design. If you ask South African officials about the origins of MeerKAT, they will point to the small endemic mongoose characteristically observed standing on its back legs (in order to see further) in the dusty plains of southern Africa. But once upon a time, when South Africa was trying to prove its willingness and ability to host the giant SKA, it built a modest seven-dish instrument, the Karoo Array Telescope (KAT-7 for short). When the designers went to the government to ask for more money, they said it was for 'meer KAT', which is 'more KAT' in the local Afrikaans language.

Today, MeerKAT dominates the arid plains of the Karoo. Like sunflowers following the Sun, its sixty-four giant metal antennas track astronomical objects and listen to the radio waves they emit. They resemble satellite TV receivers on steroids: giant white dishes, with a nose. The radio signals hit the dish, bounce off its surface to collect in the nose, and then bounce back to a receiver.

All of these dishes, which sprawl over 8km (5 miles), are connected and act as one giant telescope – just like the SKA will do. Eventually, these dishes will form the heart of the first phase of the SKA in South Africa. In total, there will be 197 dishes, many of them radiating out from the core in spiral arms and extending over about 150km (93 miles). Later phases will see more dishes in South Africa, as well as in eight partner African countries. Until then, South Africa's MeerKAT is busy day and night collecting radio signals from space.

Unlike optical astronomy, which focuses on the relatively tiny visible light range, radio telescopes collect signals from a larger part of the electromagnetic spectrum (see page 18). MeerKAT, for example, focuses on wavelengths between 3 and 30cm (1–12 inches), while visible light occupies a tiny – literally – sliver of the spectrum with wavelengths between 400 and 700 nanometres. In South Africa, SKA will collect mid-frequency signals (where wavelengths are in the centimetre range), while in Australia, its antennas will concentrate on lower radio frequencies (where we are talking about metre wavelengths).

By collecting and collating data from different radio frequencies, astronomers can map out the dark recesses beyond Earth, like blind cartographers listening to the sounds of the cosmos. Properly interpreted, the data they gather will allow them to see into the past in a way never dreamed of by early optical astronomers, and in ways previous generations of radio astronomers could not have hoped for.

OPPOSITE: An aerial photograph of MeerKAT during its construction in 2017. *South African Radio Astronomy Observatory (SARAO)*

BELOW: The MeerKAT site as seen from the air in 2018. *South African Radio Astronomy Observatory (SARAO)*

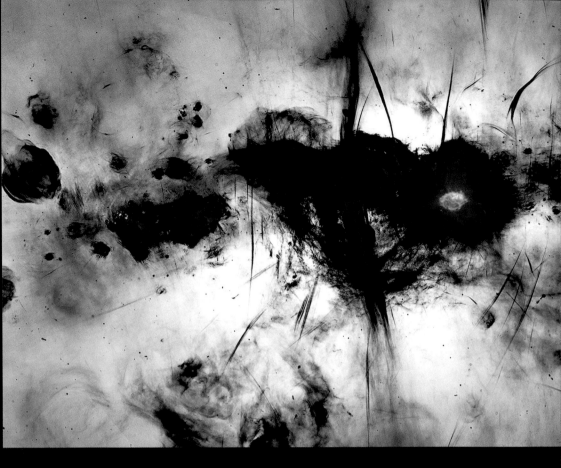

EYE OF SAURON

The MeerKAT is already giving us a new perspective on our galactic neighbourhood. In early 2022, an international team published the most detailed image of the centre of our galaxy, the Milky Way. And at its heart glows our supermassive black hole, emitting mysterious radio threads up to 150 light-years long, surrounded by the remnants of exploded stars and regions where new stars are forming. It's not possible to see such detail with your eyes, so astronomers and data scientists created an image of the black hole from the radio signals.

'When I show this image to people who might be new to radio astronomy, or otherwise unfamiliar with it, I always try to emphasise that radio imaging hasn't always been this way and what a leap forward MeerKAT really is in terms of its capabilities,' lead author Ian Heywood, an astronomer at Oxford University, said when the paper was published.

While a giant leap for radio astronomy and cosmology, the image is a bit disconcerting. It resembles the Eye of Sauron, from *The Lord of the Rings*, with angry flashes of red and grey, and a burning yellow black hole.

It is based on observations taken over about 200 hours during the telescope's commissioning phase, and the MeerKAT team has made the data available to all astronomers. 'The work we're doing with MeerKAT will eventually inform the science we're going to do with the SKA,' says Erwin de Blok, an astronomer at the Netherlands Institute of Radio Astronomy (ASTRON) and the principal investigator of a large Meer-KAT survey looking at galaxies. 'The nice thing is that, with MeerKAT, we will solve

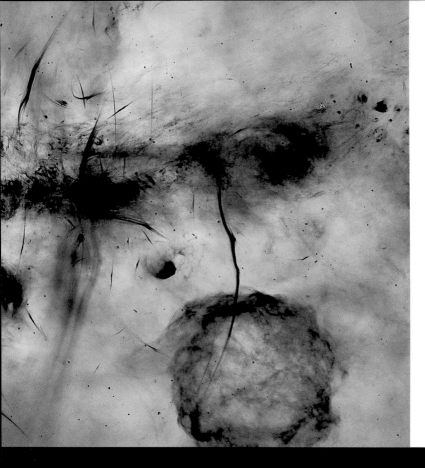

A MeerKAT image of the centre of the Milky Way. *Ian Heywood/South African Radio Astronomy Observatory (SARAO)*

some of the problems that we currently have, but we will also probably just create new questions for the SKA.'

The SKA team has set itself some lofty big-picture science aspirations, dreamed up by thousands of astronomers from around the world over the course of decades. First came the science questions – such as what the Universe looked like just after the Big Bang, as energy exploded through the void and coalesced into galaxies – and then scientists asked what kind of instrument they would need to be able to investigate them. The short answer: they would need a big one.

Such a large telescope, with its dishes and antennas so far apart, allows a sensitivity not possible with current instruments. In fact, the full SKA will be fifty times more sensitive, and be able to survey the sky 10,000 times faster, than any radio telescope array previously built. With such capabilities, it will be able to explore how magnetic fields crisscross the cosmos, holding galaxies together like butterflies in invisible cocoons, and how new stars and planets form. It may even be able to answer that age-old question that scientists have recently started to take much more seriously: are we alone in the Universe? But what is perhaps even more tantalising is the prospect of the discoveries we are not expecting.

Sarah Pearce, Director of the SKA telescope project in Australia, is excited at the thought of what such a large low-frequency array will detect, while listening to wavelengths that span metres. 'Low-frequency radio telescopes are not a new invention, but there are fewer of them around the world,' she explains. 'We will absolutely see phenomena that we wouldn't have seen otherwise.'

ERRATIC POLKA DOTS

In the next few years, the red sands and patchy khaki shrubs of Western Australia will sprout thousands of 2m (6ft) high antennas. Looking like a cross between a television aerial and a wire Christmas tree, these antennas will detect wavelengths between 1 and 6 metres (3–20ft) – significantly longer than those of the mid-frequency dishes in South Africa's dry and dusty Karoo.

In total, there will be more than 130,000 antennas, divided into arrays of 256. These array stations will spot the Australian landscape like erratic polka dots, although not so haphazardly to the radio engineers and astronomers who have spent years of their lives deciding exactly where to place them. By combining the data collected by these signal-receivers, astronomers will create the most comprehensive low-frequency map of the southern skies to date.

The first batch of antennas has been ordered – enough for six arrays of 256 – and Pearce expects that the team will start deploying them in the next couple of years. 'We have to deploy them in small arrays,'

IT ALL STARTED WITH ...

In the first moments after the Big Bang, our Universe was so hot that it was not even matter, but rapidly expanding pure energy. As it expanded from the scene of that cataclysmic explosion, the Universe cooled, creating protons, electrons and ultimately the simplest atom, hydrogen. These foundational building blocks attracted more matter, and formed the first stars, whose light filled the swirling primordial gas cloud of the early Universe. This is known as the Epoch of Reionisation. This epoch is a source of fascination for astronomers. It is when the first stars and galaxies lit up, illuminating our Universe. Over time, this radio light from primordial gas – once energetic and vivid – has relaxed, its wavelengths lengthening and shifting into lazier radio waves as it traverses the billions of years and miles that separate us from its origin.

The SKA's low-frequency array, under the clear Australian skies and far from the intrusion of towns and cities, will be able to detect primordial hydrogen gas and image this so-called 'Cosmic Dawn'.

It says something for the global ambition of the project that the SKA headquarters is on the other side of the world at the iconic Jodrell Bank Observatory in England. First used in 1945, Jodrell Bank hosts numerous telescopes, such as the massive Lovell Telescope (see overleaf). It is the hub of the project, overseeing what happens at the remote sites in Australia and South Africa.

she explains. 'We need to make sure that we don't need to change anything because changing something on every antenna when you have thousands, while not impossible, will be very difficult.'

The low-frequency array will follow some of the same science as its large-dish companion in South Africa, such as mapping galaxies and tracking their evolution, and searching for highly magnetised rotating neutron stars, called pulsars, that emit pulses of radiation. But Pearce anticipates that the Australian SKA will truly come into its own when studying the early Universe.

SPACE PIONEER

The world-famous Jodrell Bank telescope was conceived by Bernard Lovell, a pioneer of radio astronomy. Lovell was born in Bristol in 1913 and studied physics at the universities of Bristol and Manchester until the outbreak of the Second World War. During the war he helped develop radar systems for aircraft and flew on many high-altitude test missions. It was during air raids that he developed several of his scientific ideas.

After the war, Lovell acquired a small radar unit to pursue his research into cosmic rays. He was given permission to set up his equipment at Jodrell Bank, in Cheshire, and eventually got the funds to build the Mark I

"It remains an impressive feat of engineering and still uses gears from battleship gun turrets to tilt its 76m (250ft) wide dish."

telescope, then the world's largest steerable radio telescope. Since renamed the Lovell Telescope, it remains an impressive feat of engineering and still uses gears from battleship gun turrets to tilt its 76m (250ft) wide dish.

Sir Bernard's scientific discoveries include the investigation of meteor showers, cosmic rays and quasars. These extremely distant astronomical objects can emit enormous amounts of energy, including strong radio signals, X-rays and gamma rays.

In October 1957, just months after it was completed, the Mark I telescope proved to be the only way of tracking the first satellite, Sputnik 1. It also played a vital role as an 'early warning' radar to protect the UK against a surprise nuclear missile attack. Later, the telescope was used to discover gravitational lensing, which is when space-time warps around massive objects.

Lovell was knighted in 1961, for his contributions to the development of radio astronomy. In 2019, UNESCO declared Jodrell Bank Observatory a World Heritage Site, saying its 'exceptional technological ensemble illustrates the transition from traditional optical astronomy to radio astronomy (1940s to 1960s), which led to radical changes in the understanding of the Universe'.

The groundbreaking Lovell Telescope at the Jodrell Bank Observatory in Cheshire, UK.
Nigel Wilkins/Alamy Stock Photo

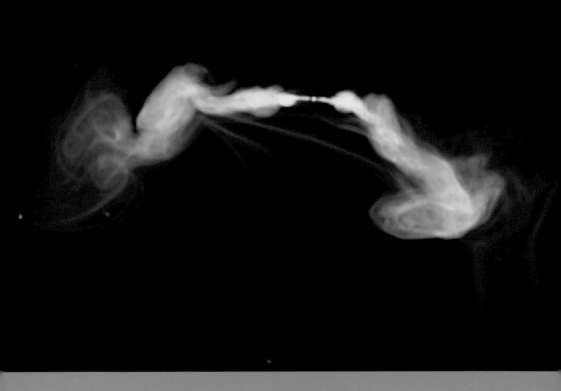

"MeerKAT produces about 2.5 terabytes of data per hour, which is more than 500 DVDs."

The SKA is also likely to radically change our understanding of the Universe. But scientists will have to be patient.

'Construction is going to take a while, and, until then, there's a lot of science to be done with MeerKAT,' says Adrian Tiplady, Deputy Managing Director of the South African Radio Astronomy Observatory. 'Already, its scientific output is having a significant impact on new discoveries and we're just scratching the surface.'

It took scientists three years to trawl through the data that yielded the latest image of the Milky Way's supermassive black hole, and that is a taste of the challenges to come for SKA radio astronomers. Gone are the days of noticing an interesting object in an image, picking up the phone, and telling everyone about it – although that may have only ever happened in movies. In modern astronomy, particularly radio astronomy, observations require trawling through rivers of data and patience. Lots of patience. MeerKAT produces about 2.5 terabytes of data per hour, which is more than 500 DVDs. But that is a drop in the ocean compared to the deluge that the SKA will bring. Every year, the volume of data will fill the equivalent of over a million 500GB hard drives.

COMPARE THE MHONGOOSE

ASTRON's De Blok is one of the astronomers sifting through MeerKAT's staggering amount of data. He leads the MHONGOOSE project, raising the naming stakes for astronomy acronyms. MHONGOOSE – which stands for MeerKAT Hydrogen (actually, a very specific form of atomic hydrogen) Observations of Nearby Galactic Objects – Observing Southern Emitters – will survey thirty galaxies to understand how they form and evolve.

Alongside MHONGOOSE, the telescope has other equally well-named surveys: 'LADUMA' (aka Looking at the Distant Universe with the MeerKAT Array) is what football fans in South Africa shout when someone scores a goal; the namers of 'MIGHTEE' (MeerKAT International Gigahertz Tiered Extragalactic Exploration) are obviously setting their survey up for greatness or, at the very least, exciting press releases; and then there's ThunderKAT (The Hunt for Dynamic and Explosive Radio Transients with MeerKAT) – which might just be the best name of all.

For many of these surveys, detecting hydrogen is vital. Hydrogen is a basic building block of stars, and galaxies need hydrogen to fuel star formation. This makes hydrogen hunting central to the MHONGOOSE project. While hydrogen is essential, there does not appear to be enough of it inside galaxies to sustain them. De Blok's project will track hydrogen's distinctive radio signal in nearby galaxies to investigate the paradox in these cosmic systems.

OPPOSITE: Threads of radio emissions, detected by MeerKAT, link different parts of the ESO 137-006 galaxy. The galaxy is about 250 million light-years away. *South African Radio Astronomy Observatory (SARAO)* **ABOVE:** An artist's representation of MeerKAT and the radio signals it can detect. *South African Radio Astronomy Observatory (SARAO)*

'This is the first time that we have an instrument that can actually address this,' says De Blok, but he warns that 'there's at least five or six years' worth of labour by many people in this survey to fully mine this fantastic data set'.

Understanding the behaviour of nearby galaxies will also feed into our knowledge of what galaxies were doing at the Cosmic Dawn, and what happened when light from those first stars and galaxies began shining out through the Universe.

CROWDED SPACE

There is, however, a threat to the science possibilities of the SKA and other telescopes around the world. A growing swarm of satellites is being launched into orbit around the Earth, blocking out or overwhelming the signals of the Universe. To optical astronomers, these moving bodies reflect light and leave white streaks across images of the night sky. To radio astronomers, they emit radio signals at wavelengths that overwhelm those given off by organic molecules, which are the

building blocks of life and important to our understanding what is out there in the vast reaches of space.

Both South Africa and Australia have national legislation protecting their SKA sites from radio frequency interference. It is illegal to use unauthorised radio-emitting devices in the vicinity of the tele-scope, because the signals could not only disrupt observations but also cause serious damage to the sensitive receiving equip-ment. Imagine listening as hard as you could to a distant bird call, only to have someone touch their lips to your ear and shout. For us, a blown eardrum; for radio telescopes, blown (expensive, custom-made) equipment.

But national governments and even international treaty organisations like the SKA Observatory have limited control over the skies overhead. The International Telecommunication Union says that nearly 100,000 satellites could be launched into low Earth orbit in the coming decade. Already, aerospace manufacturer SpaceX has launched thousands of its Starlink broadband satellites, with many more thousands planned. UK-based OneWeb has already launched a couple of hundred, while Amazon's Project Kuiper involves more than 3,200 satellites. Before long, there could very

well be a mega-swarm of satellites circling Earth, whose collective buzz and reflection threatens to deafen and blind humanity to the Universe around us.

'The SKA sites were chosen for obvious reasons,' says William Garnier, the SKA Observatory's Communications Director. 'They are protected by legislation so we can conduct our observations, but the problem is coming from these mega-satellite constellations.'

In addition to constructing a giant telescope, SKA scientists are spearheading efforts to protect Earth's skies from undue satellite interference by co-hosting a new international centre, with the RAS as a key supporter. 'There is a way we can cohabit,' says Garnier. 'What we do (as astronomers and as the SKA observatory) is also relevant because of the science, because of the innovations that will come of it, and all of the potential spin-offs.'

Without such protections, humanity's noise and technology will ripple outwards from Earth. There could come a day when even the far side of the Moon may not be far enough away to escape humanity's din. And the decades of the SKA's lifetime – and doubtless amazing discoveries ahead – are only just beginning.

"Before long, there could very well be a mega-swarm of satellites circling Earth, whose collective buzz and reflection threatens to deafen and blind humanity to the Universe around us."

SpaceX launches sixty new Starlink internet satellites into orbit on 24 March 2021. *Brandon Moser/Alamy Stock Photo*

EXPLORING THE

Two advanced NASA rovers and a helicopter are currently exploring Mars and more missions by several nations are planned in the coming years. Could we finally discover signs of life? Izzie Clarke investigates our latest efforts to explore a world that was once very different …

RED PLANET

Taking a moment to explore the twinkling night-time canvas above doesn't have to involve a fancy telescope set-up, with mounts and different eyepieces, as fun as that can be. I'm talking about simple stargazing, taking a moment to stare up at the dazzling tapestry of far-away worlds. To be quite honest, I often do it when I take the bins out or am sitting next to our living-room window (away from the cold), searching the sky for a rust-coloured dot: Mars.

I don't know how it started but I look out for Mars on most winter nights – in our flat, you 'say hello to Mars' when you spot it – and I would firmly recommend adding it to your beginner's guide to stargazing (or planet-hunting). It'll give Orion's belt a much-needed break.

In many ways, Mars can feel like a familiar planet. It is a terrain with mountains, valleys once carved out by vast rivers, and icy poles very much like our own. But the reality is that it is a complex world that we are only just starting to know.

Mars's Jezero Crater, in an image taken by the *Perseverance* rover in April 2021. The boulders in the foreground are about 50cm (20in) across on average. *NASA/ JPL-Caltech/ASU/MSSS*

CURIOSITY

'Mars is not a simple place,' says Dr Abigail Fraeman, the deputy project scientist for NASA's *Curiosity* Mars rover. 'It has a diverse history over 4 billion years that are all now tangled up in the rock record.'

It is thanks to rovers that we are beginning to scratch the surface and dig into this long and complicated history of Mars. We've already seen results from NASA's *Sojourner*, *Spirit* and *Opportunity*, which have laid the groundwork for the two current rovers operating on the Red Planet, *Curiosity* and *Perseverance*.

It is now more than a decade since *Curiosity* touched down on 5 August 2012 and, thankfully, the car-sized rover is still going strong.

'I have been involved with the mission since landing; I was a graduate student at the Jet Propulsion Laboratory [JPL] and sat at the back of the room on landing night,' says Fraeman. 'I think part of the reason it doesn't feel like ten years is because of the quality of science and the excitement of the science that we're doing.'

This exploration is all centred around the question that David Bowie was asking all along: is there life on Mars?

Well ... sort of.

By 'life' we mean tiny microbes. And, technically, the question is: was Mars ever a habitable environment?, as there is no evidence that the planet currently supports life – though plenty of intriguing signs that once it could have been possible. But I don't think that would have made Bowie his millions.

Either way, *Curiosity*'s landing site was a good place to start to unravel that mystery. Gale Crater is a unique place. The crater itself formed some 3.5 to 3.8 billion years ago when a meteor hit Mars. At 154km (96 miles) wide, Gale is also one of the deepest holes on Mars with a known watery past. It is a key ingredient in the search for (the right conditions for) life.

The initial aim was for the mission to last at least 687 Earth days, surviving at least one Martian winter in the process. But ten years on and *Curiosity* has explored ancient lakebeds, detected unexpected chemicals, survived dust storms and scaled mountains, all the while taking selfies and rock samples as it goes to help us understand our planetary neighbour.

When *Curiosity* landed, it wasted no

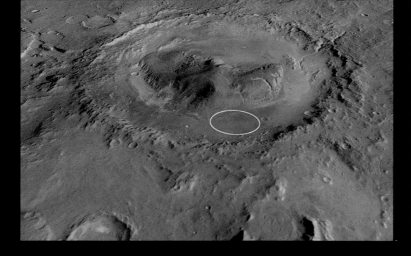

OPPOSITE: A selfie taken by the *Curiosity* rover in October 2019. *NASA/ JPL-Caltech/MSSS*

LEFT: *Curiosity*'s landing site on the Gale Crater, marked with an ellipse. For context, the marked area is 20km (12.4 miles) by 25km (15.5 miles). *NASA/ JPL-Caltech/ASU/UA*

BELOW: *Curiosity* looks back over its tracks across a Martian dune. *NASA/ JPL-Caltech/MSSS*

time in getting to work. There was a buzz in the air for the JPL team back on Earth. Fraeman remembers the excitement when a colleague shouted, 'We have images, we have images!' as they came through. Suddenly, a fuzzy, grainy, black-and-white picture was blown up on to a large screen.

Fraeman and her fellow team-mates stared at a Martian pebble. Just from a few details, they tried to pinpoint *Curiosity*'s exact location within Gale Crater – mostly in the name of science, but also partly to work out who had won the landing-site sweepstake.

DRINKING WATER

These first few images started to confirm what scientists had been hoping.

'When we looked at the rocks, we found these really rounded, large cobbles, which are the kind of rocks you expect to form in streams and rivers,' Fraeman says. *Curiosity* continued to explore the area, driving to the lowest point in the crater where, if there had been a river as they'd anticipated, a lake would have formed.

'When we got there, lo and behold, we found what were definitive lakebed sediments. When we sampled the composition of those lake rocks, we found that it was really rich in clay minerals. It had compositional indicators that suggest that the water would have been drinkable and, combined all together, it would have been a really good place for ancient life to have taken hold if it ever existed.'

It was a good start for the rover. Although *Curiosity* might look like WALL-E's older cousin, it has far more to it. It is effectively a laboratory on wheels, carrying ten different instruments. There are multiple cameras in the mast and arm for studying the Martian terrain and rocks at different scales, as well as spectrometers for analysing their chemistry. There is a small drill in the arm to take rock powder samples and feed them into two different instruments inside the rover – one to find their chemical composition and the other for calculating mineral structures, both with great preci-

"They're the kind of rocks you'd expect to form in streams and rivers."

sion. Plus, the rover even has a laser that zaps rocks to look at the plasma created. This reveals more about their composition.

And it's not just about the rocks. The environmental detectors have their own weather station, which has been particularly helpful during dust storms. These are the conditions that have put other rovers on their last legs (or wheels), but not *Curiosity*.

This rover charges in and, true to its name, studies what is going on.

Finally, there is the radiation detector. Mars is a hostile planet; we puny humans would be no match for the radiation found there. But this is the piece of kit that will help us better understand what astronauts might encounter when (or if?) they land on Mars.

COW BURPS

One of the biggest mysteries of recent Mars exploration is *Curiosity*'s detection of methane. Livestock produce a significant amount of methane here on Earth but, at last check, no Martian cattle have been spotted wandering across the red desert. But why get excited over methane? Cows may burp this odourless gas on Earth, but methane is actually produced by microbes to help livestock digest plants. So, could this mean that there are microbes living on the Red Planet?

It's a head-scratcher.

One suggestion is that methane could be produced by geological processes between rocks, water and heat. But here's another puzzle: while

OPPOSITE: The Martian pebbles, or 'blueberries', that help provide evidence of Mars's ancient watery environment. The area shown is 3cm (1.2in) across. *NASA/JPL-Caltech/Cornell/USGS*

BELOW: Strata at the base of Mount Sharp. The dip in the foreground indicates there was once a flow of water towards a basin that existed before the mountain formed. *NASA/JPL-Caltech*

Curiosity has continuously detected methane at the surface of Gale Crater, ESA's Trace Gas Orbiter (TGO) – an orbiter that, you guessed it, detects small concentrations of gas – hasn't found any methane higher in the Martian atmosphere. And we are talking small traces here.

NASA described *Curiosity*'s methane measurement as 'equivalent to about a pinch of salt diluted in an Olympic-size swimming pool', with 'baffling spikes' that raise this to roughly forty pinches of salt.

In July 2021, scientists even released a paper that explored all the possibilities of *Curiosity* itself creating this signal. It wasn't. The latest proposal is that perhaps both TGO and *Curiosity* could be right. TGO needs sunlight to detect a signal above the surface. What if, during the day, the methane signal is dispersed across the atmosphere, making it too diluted for TGO to detect? Then at night, without the sunlight to churn up the atmosphere, that signal becomes concentrated at the surface. Another option is that perhaps something is destroying methane quicker than expected. Either way, the search for its source continues – and, meanwhile, *Curiosity* takes the high road. Quite literally.

Rising from the crater floor, Mount Sharp has always been a key target for the mission. This rocky mountain is 5.5km (3.4 miles) high – it's three times higher than the Grand Canyon is deep – and *Curiosity* is measuring it layer by layer to understand the planet's history. This year, it has been studying a transitional area that contains an insight into the global changes in Mars's climate about 3 billion years ago. But the next year is all about moving into a new era.

'We'll get into a whole new phase of the mountain where we're going to get to look at a very different period in Mars's history and ask questions about what habitability on the planet was like at that time. Did it persist, or did it disappear?' says Fraeman. 'And what were the other consequences of this global planet change as it started to dry up?'

It's like if you were hiking up the Grand Canyon, starting from the bottom and working your way to the top. No doubt, you'd probably be very out of breath, but you'd also notice changes as you climbed through its geological history. The higher you go, the younger the rock layers are. You'd be walking forward through time and that's exactly what Fraeman is doing on Mars with *Curiosity*.

But there's also a new rover in town.

OPPOSITE: This view from the Mastcam on *Curiosity* shows an outcrop within the 'Murray Buttes' region on lower Mount Sharp. *NASA/JPL-Caltech/MSSS*

BELOW: A view of a field of dunes on lower Mount Sharp, taken by *Curiosity*'s Mastcam. *NASA/ JPL-Caltech/MSSS*

"What were the other consequences of this global planet change as Mars started to dry up?"

PERSEVERANCE

'*Perseverance* was originally billed as a build-to-print copy of *Curiosity*,' says planetary scientist Emily Lakdawalla, author of *The Design and Engineering of Curiosity*. NASA's *Perseverance* rover, lovingly known as 'Percy', is the next step in understanding the Martian landscape. Living up to its name, *Perseveranc*e launched on 30 July 2020, negotiating the challenges of travelling to another planet during a global pandemic and successfully arriving there in one piece. It landed in Jezero Crater on 18 February 2021 to 'follow the water' and look for that elusive ancient life, but also to collect and cache samples that will one day be returned to Earth.

There are aspects of the *Perseverance* rover that are identical to *Curiosity*. A lot of its internal electronics are the same: its buggy suspension system to traverse the rocky plains; the size and shape of the mast in the body; how the system moves heat from the power supply into the body of the rover to keep everything nice and warm. We've seen it before.

'But it has a completely different set of instruments,' says Lakdawalla. 'And so that makes it a very different mission.'

Curiosity is all about studying Mars at its surface and it has everything it could possibly need to do so. *Perseverance*, however, is there to document the planet more deeply: to take images, to drill into rock cores and collect samples, and, most importantly, to store these samples in sealed tubes that will be left on the Martian surface.

The rover has a large 'hand', officially called a turret, which houses a few instruments and the crucial coring drill that bores into Martian rocks. This arm then rises towards the body where the samples are dispensed into titanium tubes and kept safely inside the rover to be dropped on Martian soil at a later date. It'll be down to a future mission to retrieve the tubes and return them to Earth.

Another skill that *Perseverance* has is its ability to take selfies. Again, this is mostly for vital engineering health checks, but who wouldn't want the odd holiday snap from a rover roaming another planet?!

"Who wouldn't want the odd holiday snap from a rover roaming another planet?"

OPPOSITE: A selfie taken by the *Perseverance* rover in April 2021, with the *Ingenuity* helicopter in the background. *NASA/ JPL-Caltech/MSSS*

BELOW: *Perseverance*'s backshell on the surface of Jezero Crater in an image taken by *Ingenuity* in April 2022. The tangle of cables on the top of the backshell are high-strength suspension lines that connect the main structure to its parachute (upper left). *NASA/JPL-Caltech*

It hasn't always been smooth sailing on Mars. Percy's first attempt to take samples did not go quite to plan when, after drilling down into the Martian surface in August 2021, no rock or dirt actually made it into the sampling tube. A hole in one? Not quite. Fortunately, Percy is carrying forty-three sampling tubes, thirty of which NASA hopes to fill, so there were other chances. Its next attempt in September 2021 was far more successful. *Perseverance* unfurled its arm, placed its drill at the surface, dug down about 6cm (2in) to extract a rock core, and sealed it in a tube. This was the first time a rover had collected a sample that could make a return to our own planet.

Frankly, there have been many 'firsts' for this mission. None was more techno-logically astounding than the first flight of Percy's sidekick, *Ingenuity* (aka 'Ginny'). The small helicopter weighs only 1.8kg (4lb) and NASA has big hopes for a future of flying drones on the Red Planet.

When Ginny first took flight on 19 April 2021, it proved that JPL had cre-ated an aircraft that is lightweight, is able to generate lift in Mars's thin atmosphere and can survive the planet's bitingly cold nights of –90°C (–130°F). In the future, NASA aims to use similar devices to help explore Martian terrain, reaching areas that are simply too difficult for rovers. Here's hoping they call the next helicopter *Rondezvous* – aka 'Ron' – so that, along with Percy and Ginny, the Weasley/NASA legacy can continue on the Red Planet.

That said, the US isn't the only country covering new ground on Mars. Just ask the China National Space Administration (CNSA) and the Emirates Mars Mission.

CHINA'S HEAVENLY ROVER

Let's start with China. With a string of successful lunar exploration missions already under its belt, CNSA set Mars as its next target. And it was an ambitious one. The Tianwen-1 mission, meaning 'questions to heaven', has sent an orbiter, lander and rover all at once. But the risks and challenges have paid off and China is now seen as a major player when it comes to space exploration. They are, after all, the only other country aside from the United States that has reached the Martian surface.

The mission launched on 23 July 2020 and, while *Perseverance* was landing on Mars in February 2021, Tianwen-1 entered orbit and started mapping the planet from pole to pole. CNSA has set out an exciting list of objectives – which are just as ambitious as packaging up three robots and sending them to another planet on the agency's first attempt. They range from mapping Mars's geological structure and measuring the Martian climate and surface to investigating soil characteristics and water-ice distribution, as well as studying invisible features like Mars's gravitational and electromagnetic fields. Safe to say, Tianwen-1 has it covered.

Zhurong, the rover, touched down in May 2021, overcoming its own risky 'nine minutes of terror', to join the search for life by tracing a suspected ancient coastline in Mars's northern hemisphere.

Then, joining in with the July-launch madness, the United Arab Emirates inaugurated their mission, an orbiter named *Hope*.

OPPOSITE TOP: NASA's *Ingenuity* helicopter, in an image taken by the Mastcam on the *Perseverance* rover in June 2021. *NASA/JPL-Caltech/ASU/MSSS*

OPPOSITE BELOW: *Perseverance*'s first attempt to collect a sample from a hole drilled into the Martian rock, August 2021. *NASA/JPL-Caltech*

BELOW: An image released by CNSA on 1 January 2022, taken by the Tianwen-1 Mars mission, showing the orbiter flying around the Red Planet. The orbiter's picture was taken by a camera released by the craft. *Xinhua/Alamy*

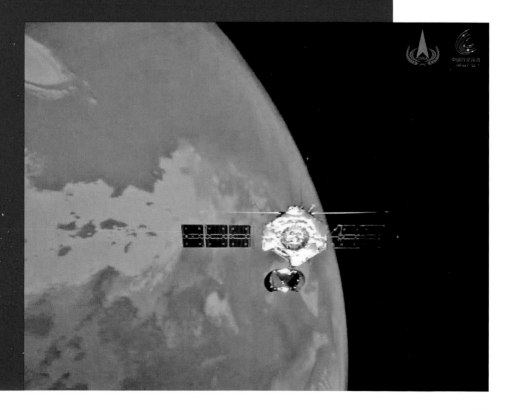

"Yes, there was a risk, but we didn't want to waste that chance."

Now, you might be wondering: why did everyone launch in July 2020? It's a good question. Personally, it was a great extended birthday present for me. But, scientifically speaking, it's all because Mars and Earth were aligned, which happens about every two years. If any of the agencies missed this two-week window to launch, they'd be looking at a twenty-six-month wait for their next opportunity.

'It was 15 July 2020, sixteen hours before launch, and Mother Nature imposed a challenge on us,' says Omran Sharaf, Project Director of the Emirates Mars Mission. 'There was a cloud layer up in the atmosphere and it was risky for us to launch. The difference in temperature could have caused issues with the launch systems.'

After two days of waiting, the team were ready for a second attempt. But Mother Nature struck again. Cue two more days of waiting, and on 19 July 2020 Omran decided to take a risk.

'That launch window started shrinking. We looked at all the weather patterns and I made the final call two hours before [launch]. The weather was a little bit funny and literally right after the launch, the weather turned upside down. Yes, there was a risk, but we didn't want to waste that chance.'

And, with that, the *Hope* probe began its journey to provide a complete picture of the Martian atmosphere and its layers, reaching the Red Planet in February 2021.

One of its landmark moments came in April 2022 with the discovery of a worm-like aurora. It extends thousands of kilometres, starting from the planet's north pole and travelling halfway down. Aurorae are a colourful display that result from the Sun's charged particles hitting a planet's magnetic field, like the aurora borealis (northern lights) here on Earth.

But, the thing is, Mars doesn't have a global magnetic field, only pockets of localised magnetic fields at its crust. And so this latest discovery has scientists returning to the drawing board with questions about the interactions between Mars's atmosphere, the planet's magnetic field and the solar wind.

But for the UAE – a newer name in the space exploration game – this mission aimed to tear up the rule book. The *Hope* probe is an incredible feat of engineering, weighing 1,350kg (2,976lb) and about the size of an SUV. Plus, it is no small achievement that, without the heritage or the budget of other agencies, the Emirates Mars Mission designed, launched and sent a probe to Mars in just seven years. Their creative risks have been worth it.

A rocket carrying *Hope*, a Mars orbiter developed by the United Arab Emirates, takes off from Tanegashima Space Center in southwestern Japan in July 2020. *Newscom/Alamy Live News*

INTO THE UNKNOWN

But what does the future hold for subsequent missions to Mars?

ESA had hoped to make the most of the next launch window in September 2022. Their UK-built *ExoMars* rover was intended to be the first rover to drill 2m (6.5ft) into the Martian surface. However, as sanctions were imposed on Russia due to the war in Ukraine, ESA suspended the launch in March 2022, with their ruling council citing 'the impossibility of carrying out the ongoing cooperation with Roscosmos'.

Then, of course, someone needs to pick up *Perseverance*'s cached samples.

Mars rovers might have a brilliant suite of instruments on board but they're simply no match for the advanced laboratories and techniques back on Earth. But how do we courier small tubes of rocks and soil between two planets?

Well, first up, good ol' Percy has samples stored internally as well as along the surface of Mars. But there are two more, huge, steps to go and ESA is working with NASA to make the leap together by 2030.

Step two will involve sending a Sample Return Lander to the Gale Crater to retrieve Percy's samples. On board are two small helicopters for backup, which I suggest we call Fred and George, to grab the samples in case *Perseverance* fails. A

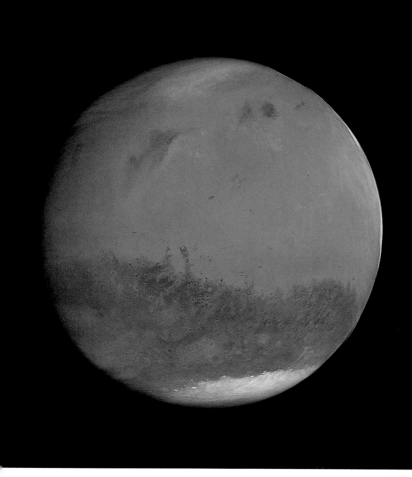

Sample Transfer Arm will take those tubes and put them into a container the size of a basketball. This will be loaded into another element, a Mars Ascent Vehicle, which is a fancy term for 'rocket'. That rocket will (somehow) blast off from surface, into Mars's orbit, and release the basketball container.

Step three will be in the form of an orbiter around Mars, waiting to return to Earth. It will time its orbit with the sample container, capture it, and speed on back home with the samples in tow.

See what I mean? This really is rocket science. Extremely difficult rocket science. But will the future of exploring the Red Planet involve humans?

'Gosh, I hope so,' says Fraeman. 'Humans to Mars for the sake of exploration,

to do field geology, could probably do what *Curiosity* did in ten years in a month or two.'

But Emily Lakdawalla has a different opinion.

'I'm not so interested in that because once you put humans on Mars, then it becomes really hard to find Martian life. Earth life is just really good at taking over. I'd much rather have robotic exploration combined with humans in a way that doesn't actually put humans on the surface.'

So, the jury is still out. But whether the future of exploring Mars involves humans, returning robots, or a mixture of the two, one thing is certain: over the next few years, we will get to know Mars like never before.

HIDDEN FIGURES OF ASTRONOMY

Sue Nelson investigates some of the unsung women heroes of astronomy.

O ver 1,600 years ago, an astronomer, like Beyoncé, was known by one name: Hypatia.

A philosopher and mathematician as well as an astronomer, Hypatia might not have had the mass appeal of Beyoncé, but in her own way she was equally famous. Admired for both her brains and her beauty, she would lecture about the stars at the famous Museum of Alexandria in Egypt, attracting audiences from distant lands.

Just like Beyoncé, Hypatia's father played an instrumental role in guiding his daughter's career. Theon was a Greek scholar and mathematician from a wealthy family and he ensured that Hypatia was extremely well educated. For a girl during the fourth century AD, this was unusual, to say the least. Hypatia never married and had an incredibly successful career. In short, she was the ultimate single lady.

According to the historian Philostorgius, a contemporary of Hypatia, when it came to observing stars, her talents surpassed even those of her father. 'Hypatia had discovered something new about the motion of the stars and she made this new knowledge accessible to men and women of her time,' he wrote, 'explaining her new observations in an original work, which she entitled *Astronomical Canon*.'

It's a tantalising prospect but sadly there are no surviving copies of this canon – a collection of tables about celestial bodies – and so her work and discoveries remain unseen. We do know that, at the age of thirty-one, Hypatia succeeded her father as a teacher and mentor at the museum. Nepotism aside, it's an astonishing reflection of her intellectual capabilities that she was trusted to lead others at this ancient centre of classical learning.

However, her fame came at a price, and in AD 415 she was assassinated, most likely by religious fanatics. Fortunately, her name lives on even if, for many, it is hidden in plain sight. Look upwards during the next full Moon: one of those craters is named Hypatia.

Many believe the feminine-looking figure peering out from the crowd in Raphael's masterpiece is the philosopher and astronomer Hypatia of Alexandria. *Detail from* The School of Athens *by Raphael, 1509–1511. Album/Alamy*

PRINCESS ENHEDUANNA

There are records of other, even earlier female astronomers – such as Aglaonice (also known as Aganice), Theano (believed to be the wife of the Greek mathematician Pythagoras) and Fatima of Madrid – though far less is known about them. But the award for best job title has to go to Princess Enheduanna. In the Sumerian region of Mesopotamia (modern-day Iraq) around 2300 BC, she was high priestess of the Moon goddess in the city of Ur, referenced in the Bible's Book of Genesis as the birthplace of Abraham. Her role included studying the movements of the stars and the Earth's Moon.

Today the royal priestess is best remembered as a poet and is often referred to as the first writer recorded in history by name. But Enheduanna's connection to astronomy also endures. In 2015, a crater on the planet Mercury was named after her.

As it happens, crater names are a great resource for discovering interesting female astronomers, many of whom history has long forgotten. The far side of the Moon, for instance, is an apt place for a crater named after another woman few people know about.

> "The award for best job title has to go to Princess Enheduanna ... high priestess of the Moon goddess"

LADY COMPUTER

Technically speaking, the Maunder crater represents both Irish astronomer Annie Maunder and her astronomer husband Edward Walter Maunder (known as Walter). Even during Annie's life, the full extent of her contributions was often overlooked.

The Heavens and Their Story, a popular astronomy book first published in 1908, featured both their names on the cover even though the preface by her husband states: 'The present book, which stands in the joint names of my wife and myself, is almost wholly the work of my wife, as circumstances prevented my taking any further part in it soon after it was commenced.'

Admittedly, it was probably the publisher's decision to give Walter Maunder equal billing, because he was a known name from working at the Royal Observatory at Greenwich in London. This is where Annie met her future husband and began specialising in an area of astronomy in which she would make a number of important contributions.

Annie (born Annie Scott Dill Russell) had studied mathematics at the University of Cambridge, sitting her final exams in 1889 at a time when degrees were not awarded to women. Two years later she began work as a 'lady computer' in the solar department at the Royal Observatory.

Since laptops had not yet been invented, a 'computer' meant something different in those days. It referred to someone who made mathematical computations on paper – remember, there were no calculators around either – and, though it required a sharp brain, it was often a poorly paid job.

Annie's job involved analysing daily photographs of the Sun. It led to a lifetime's interest in solar activity and sunspots, although she was forced to resign in 1895 because civil service rules didn't allow married women to work. However, that didn't stop Annie's research.

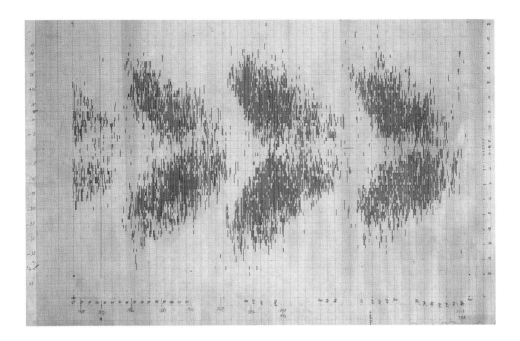

BUTTERFLY SUN

'There are some really important articles in the history of solar physics,' says Sian Prosser, the Royal Astronomical Society (RAS) archivist. 'For example, the butterfly diagram, where it's clear that Annie Maunder collaborated in crunching the data that went into making it, showing the position of sunspots over the centuries.'

Plotting the latitude of where sunspots occur over time produced a series of shapes resembling butterfly wings, which gave the diagram its name. These patterns showed that solar activity has a cyclical nature: when a new solar cycle starts, for instance, sunspots tend to form at mid-latitudes and then move towards the equator.

It was an important piece of work and the graph was submitted to a meeting of the RAS in 1904 – still twelve years before the society admitted women. Annie Maunder's name often appeared as co-author with her husband on scientific papers but, for some reason, not this one. 'It's just Edward Walter Maunder's name on the actual article itself,' says Prosser.

Yet, as even her husband confirmed on numerous occasions, Annie contributed to his work. In the same year, Annie wrote in a letter: 'We have been working very hard this summer on the connection between sunspot activity and the magnetic storms ... While my husband is in Ireland, I am working.'

The connection Annie and Walter made between geomagnetic activity and sunspots is a crucial feature of understanding space weather. Sadly, many textbooks and scientific papers ignored her contributions.

'Another thing that I've noticed in the archives,' says Prosser, 'is that we don't have much under her name although there are loads of letters written to the RAS under her husband's name – but it's in her handwriting.'

Annie was also the first editor of the British Astronomical Association journal and took part in numerous eclipse expeditions. Happily, her name is now emerging – aptly enough considering her field – into the sunlight.

In 2016, the RAS established the Annie Maunder medal for an outstanding contribution to outreach and public engagement in astronomy or geophysics. More recently, in March 2022, English Heritage commemorated both Annie and Walter with a Heritage London blue plaque at 69 Tyrwhitt Road in Lewisham, London, where they once lived. This is where they (or mostly Annie) wrote *The Heavens and Their Story*.

PAINSTAKING WOMEN AND IMPATIENT MEN

Annie wasn't the only 'lady' human computer who turned an often repetitive and painstaking job into a career that advanced our knowledge of astronomy.

In the United States, from the end of the nineteenth century until the 1940s, the female Harvard 'computers' analysed the spectra of stars on glass photographic plates at the Harvard College Observatory in Cambridge, Massachusetts. Spectra (plural of the Latin word *spectrum*) refers to the light collected from stars over a range of intensities.

Isaac Newton had discovered several hundred years earlier that a prism can disperse white light into a rainbow. This spectrum of light is a source of information since it reveals the different radiation energies and colours of the visible part of the electromagnetic spectrum for the human eye. For example, we see light waves that are around 650 nanometres in length – that's 0.000000650 metres – as the colour red.

Astronomers discovered that if you looked at the stars with a prism in front of the telescope, the same thing happened.

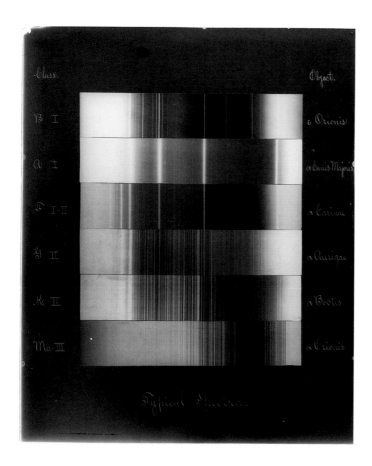

OPPOSITE: Annie Maunder's butterfly diagram. *© 2022 UCAR*

LEFT: A spectral plate showing the tiny smudges that had to be painstakingly analysed by the so-called Harvard 'computers'. *Harvard College Observatory, Astronomical Glass Plate Collection*

The stars' distant light would disperse into small spectra and these could then be captured on glass photographic plates.

Having seen some of these plates at first hand, I can testify to the skill and scrutiny this job must have required. The spectra are often tiny, 1cm (0.4in) long grey smudges – like a smear of pencil on glass (the plates are negatives so the brighter the star, the larger the smear). It is only under a magnifying glass that these smears reveal vertical lines like a barcode. The position of these spectral lines corresponds to the activity of specific elements within the stars – such as hydrogen or iron – and tells you about their chemical make-up. Spectra can also provide information on temperature and density.

Originally, the job of a computer was done by men. The reason for employing women was primarily economical. 'Not only are women available at smaller salaries than are men,' wrote astronomer Frank Schlesinger in 1901, 'but for routine work they have important advantages. Men are more likely to grow impatient after the novelty of the work has worn off and would be harder to retain for that reason.'

There are around a million of these glass plates, making up a collection of 'plate stacks' in brown paper covers, rather like vinyl record sleeves. The plates and observations continue to be used for research today, since they have captured over 500,000 celestial moments going back well over 100 years.

It is only relatively recently that the importance of the women's work has been recognised. Several of the computers – including Annie Jump Cannon, Henrietta Swan Leavitt, Williamina Fleming and Antonia Maury – became respected astronomers in their own right and are also memorialised by lunar craters. Although, rather like Annie Maunder, Antonia doesn't get a crater of her own and shares her glory with a relative – in this case her cousin, Lieutenant Matthew Fontaine Maury from the US Naval Observatory, whose name was first bestowed on the crater.

Antonia Maury originally joined the Harvard College Observatory in 1887 to work on the Draper Memorial Project. This catalogue of stars was being produced to honour her late uncle, Henry Draper – a physician, amateur astronomer and keen astrophotographer – and was funded by his widow.

Astronomers will be familiar with the Hertzsprung–Russell diagram, which is still used today. It was developed in the 1900s and classifies stars according to their brightness and colour. Years earlier, however, Maury had already spotted this connection – that some stars with the same brightness had different colours – and had sorted stars into these different categories while working

on the Draper Memorial Project.

Feistier than most, Maury refused to be overlooked and insisted on being acknowledged as the author of her own research. The Draper star catalogue, the first under a woman's name, was published in 1897 as 'Discussed by Antonia C. Maury under the direction of Edward Charles Pickering'.

Danish astronomer Ejnar Hertzsprung then used Maury's observations in this catalogue as a reference for his work. At the same time, an American astronomer, Henry Norris Russell, independently came up with the same diagram, which is why it's called Hertzsprung–Russell. A more accurate name, however, would be Maury–Hertzsprung–Russell.

Then there was British-born Cecilia Payne (later Payne-Gaposchkin) who arrived at the observatory in 1923 as a research scientist and so was not technically a computer. But in common with the computers, she was poorly paid.

Two years later, Payne became the first person to get a doctorate in astronomy from Harvard. She discovered while doing her thesis that stars are made up of mostly hydrogen and then helium – something we take for granted today – and could be classified by temperature. Her supervisor, also the observatory's director, as well as astronomer Henry Norris Russell, both disagreed. Payne was right. Payne-Gaposchkin's name may not be on the Moon but it is on both an asteroid and a volcanic crater on Venus.

"Feistier than most, Maury refused to be overlooked and insisted on being acknowledged as the author of her own research."

SCOTTISH MAID

Williamina Fleming was definitely another Harvard computer and her work is finally receiving public acclaim. She discovered the extremely hot, dense 'white dwarf' stars and was the first person to see the Horsehead Nebula beyond the eastern star of the belt in the constellation Orion – admittedly on a photographic plate and not in glorious technicolour like the images produced from the Hubble Space Telescope. She also classified over 10,000 stars, something astronomer Herbert Hall Turner deemed 'an achievement bordering on the marvellous'.

Originally from Scotland, Fleming worked as a teacher before emigrating to Boston with her husband in 1878. When Fleming was pregnant with their first child, he abandoned her, forcing Fleming to seek a position at the observatory as a maid. Fortunately, the observatory director at the time, Edward Pickering,

recognised her intelligence and talents and she soon switched from dusting the furniture to working as a computer. She would go on to become the observatory's first Curator of Astronomical Photographs and one of America's most prominent female astronomers. In 1906 Fleming became the first American woman elected as an honorary member of the RAS.

Born a few years after Fleming, in 1863, Annie Jump Cannon built on Fleming's work and became a renowned expert in classifying stars. She also combined two stellar classification methods – a joint one by Fleming and Pickering and the other by Antonia Maury. The Fleming–Pickering method was an alphabetical one dividing stars into twenty-two classes dependent on hydrogen content – A-type stars containing the most hydrogen. Combining the two methods meant these spectral classes, which are still in use today, no longer follow an A, B, C order, but instead run as follows: O, B, A, F, G, K, M. This used

"Fleming was the first person to see the Horsehead Nebula."

now known as Leavitt's Law. She worked out how to measure the 'intrinsic' luminosity of stars – how bright a star is as opposed to how bright it appears to be, an important distinction. Otherwise, two stars of equal brightness could appear to be the same distance from Earth when one might be a huge bright star far away and the other a smaller, less bright star but a lot nearer.

Leavitt's insights enabled astronomers to measure the distances of remote stars and galaxies and proved crucial for Edwin Hubble to prove that the Andromeda galaxy was outside our own. A testimony to the importance of Leavitt's work is that a member of the Swedish Academy of Sciences nominated her for a Nobel Prize in 1924. Unfortunately, the member hadn't realised Leavitt had died of cancer a few years beforehand. The prize is not awarded posthumously.

to be recalled by the mnemonic 'Oh Be A Fine Girl and Kiss Me', though nowadays #MeToo may have put paid to that. Jump Cannon classified around 350,000 stars – an extraordinary feat – and was the first woman ever to receive an honorary degree from Oxford University.

Then there was Henrietta Swan Leavitt, who identified that the luminosity of a variable star is related to its pulsation period,

All these achievements are even more impressive when you realise that many of the women were not allowed to do their own observing as part of their day jobs at the observatory, meaning that their insights and research were primarily done outside work and on their own time.

RAZOR-SHARP HOFFLEIT

In 2004, I was lucky enough to meet a woman who had actually worked along-side some of these computers. Dorrit Hoffleit, like Cecilia Payne, was never one herself. She had a degree from a women's college and joined the observatory, in 1924, as a research assistant to Henrietta Scope, who studied variable stars.

Ellen Dorrit Hoffleit was ninety-six when I took a train from Boston to meet her at New Haven, Connecticut. She sparked with vitality, had a razor-sharp mind, was quick to giggle and told me how she first became interested in astronomy.

'When I was eleven years old, my mother and I were out in the back watching the Perseid meteors and what happened shouldn't have happened. One Perseid was coming in like this ...' – she raised her right hand – 'and another was coming in ...' – she raised her left hand. 'And they collided. I didn't dare tell anyone about this afterwards because I thought they'd say I was hallucinating!'

One meteor specialist wrote that the chance of a collision occurring like this was once every 30 million years. No wonder Hoffleit decided to later make shooting stars the subject of her life's research, along with variable stars – those whose brightness fluctuates over time. We spoke in her office at Yale University, where she worked and wrote astronomical research papers until her death, at the age of 100, in 2007.

Hoffleit recalled Jump Cannon as having 'a charming personality. Everyone liked her.' If she didn't want to prolong a heated discussion with the observatory director, Harlow Shapley, 'Miss Cannon, who was hard of hearing, unplugged her hearing aid and wouldn't listen to any more that Shapley had to say,' Hoffleit claimed.

Hoffleit remembered her time working at the observatory alongside the computers as 'fun', and like the other women she was proud to be there. She praised Maury's contributions, in particular, as being 'a great achievement'. However, she was understandably indignant about the structure of academic astronomy during that era. 'You appoint a woman,' she said, 'and then you let her do the work and be the co-author of what she'd already done!' It seems not much had changed since Annie Maunder's day.

THE FLAME OF FIAMMETTA WILSON

Hoffleit wasn't the only pioneering woman who found shooting stars utterly absorbing. Born over forty years before her, an amateur British astronomer, Fiammetta Wilson, also loved observing meteors.

'Fiammetta recorded over 10,000 meteor observations,' says Bill Barton, Deputy Director of the British Astronomical Asso-

ciation Historical Section. 'Over 650 were doubly observed and a doubly observed meteor is much more useful for science as you can triangulate its path through the Earth's atmosphere.'

Wilson was a fascinating character. One of the first female Fellows of the RAS, she was a meteor observer for the British Astronomical Association (BAA), as well as speaking several languages and studying music.

She had arrived into the world in Suffolk as Helen Frances Worthington, in 1864, the eldest of five children. Wilson was her married name from her second husband and the name 'Fiammetta' made an appearance after her first marriage had ended. At one point she was also known as a Russian professor from Poland under the name Fiammetta Waldahoff, despite being neither a professor, nor Russian, nor from Poland. But she did teach mandolin at London's Guildhall School of Music.

'She was basically abandoned by her first husband within a couple of years of their marriage,' says Barton. 'He'd committed adultery so Fiammetta had to get money from somewhere and fell back on what she knew.

'As a child she'd had a home governess, had been to finishing school in Germany and spent a year in Italy studying music,' Barton adds. 'Waldahoff appears to be a complete invention – it just appears in the 1901 census.'

As for the intriguing adoption of Fiammetta as a new name, it is the Italian word for 'little flame' and a marvellously apt choice for a woman who loved observing meteors. But there's another connection too. 'The name Fiammetta comes from Italian literature,' says Barton, 'and it is a lady who is extremely poorly treated by men.'

Wilson's interest in astronomy began after hearing astrophysicist Professor Alfred Fowler give a lecture at the Imperial College of Science and Technology in London, around 1910. She would go on to make her own observations from the back garden of her marital home in Hertfordshire, using binoculars and, later, a telescope, and there she discovered her passion for meteors.

During the First World War, Wilson became a temporary co-director of the BAA's meteor section with Alice Grace Cook and, while Wilson never made a significant discovery, Barton believes her name should be remembered because of her prolific observations.

Fiammetta Wilson.
Courtesy of The Royal
Astronomical Society
Library and Archives

SHINING STARS

Wilson's co-director, known by her second name Grace and thirteen years Wilson's junior, was another amateur astronomer whose passion and enthusiasm for observations and classifications also contributed to the field.

Born in Stowmarket, Suffolk, in 1877, Cook's family was wealthy enough for her not to work and to devote herself to stargazing. Cook classified 785 objects on British astronomer John Franklin-Adams' photographs of the night sky, taken between 1902 and 1910. 'These images captured not only stars, but nebulae. At this point in history, many of these objects remained mysterious,' says Barton. 'So she sees small objects, diffuse objects, and creates a category for spiral objects, which we now know to be spiral galaxies.'

Photographs of Cook at her home show a surprisingly large observatory in her back garden. This is because the BAA, which was the only astronomical society that allowed women to join at that time, once had a collection of instruments – from small telescopes to observatories – that members could borrow. 'You applied for an instrument and it was granted to you,' says Barton. 'You somehow got it delivered and then you would need to write annual reports to show that you were using the instrument and it was in good working order.'

Cook discovered an unknown meteor stream and was the first person in the UK to see the Nova Aquilae (or V603 Aquilae) in 1918. It was the brightest object of its type throughout the twentieth century and could be seen in the night sky with the naked eye for several months. In 2000, Nova Aquilae was confirmed as a binary star – one of two stars that orbit each other.

"Photographs of Cook at her home show a surprisingly large observatory in her back garden."

Apart from observing meteors, Cook enjoyed studying the aurora borealis, or northern lights. There's a lovely connection with the 'lady computers' too as, in 1920, the Harvard College Observatory gave her a grant of $500 for her astronomy work.

There is no doubt that there are many more names to uncover of women who have played an important role in humanity's oldest science: observing the stars. Some, like the eighteenth-century German astronomer Caroline Herschel, are finally receiving interest in their own right. In the past ten years alone, at least six books have been published either devoted to her on her own or combining her achievements alongside those of her famous brother, William. For many years she had just been seen as the scribe of William's astronomical notes. But even her note-taking made a difference.

'Nearly everything that we have in the archives is written by her,' says Prosser at the RAS, 'and I have seen how she expands and clarifies on his observations. In fact, we have William's original contemporary notes from the night that he discovered Uranus and then, nearly forty years later, he asked her to write up all of his observations and put them in a set of notebooks. And because she was there at the time, and because she knows what her brother means, she will make what he is doing more comprehensible. So she's having that influence on the perception of his work.'

Women's contributions to astronomy, in all their forms, hidden or not, are coming to the fore. The crater on the Moon, C Herschel, celebrates Caroline in her own right and it's clear there are many more women whose names will one day become a feature on another planet or make it on to the lunar surface.

As it happens, the astronomer who was responsible for sorting out a consistent nomenclature for the Moon's features was also a woman. Mary Adela Blagg, another pioneer among that first batch of female RAS Fellows in 1916, mapped the Moon and – at a time when one crater could often have several names – created a clear system out of impending chaos.

What I love about all these different women is that they followed their passion. Whether they were a single mother in the nineteenth century or an educated priestess in what was very much a man's world, these women were stars in so many ways and their legacy continues to shine.

Grace Cook in her back-garden observatory. *From* The English Mechanic and World of Science *issue 2604 (19 Feb 2015)/ courtesy of The Royal Astronomical Society Library and Archives*

JUICE TO

Sarah Wild looks forward to the launch of an ambitious mission to explore Jupiter's icy moons. These mysterious worlds could be our best chance of finding life elsewhere in the Solar System.

JUPITER

In early summer 2023, Professor Michele Dougherty will have one of the most nerve-racking hours of her life. She's responsible for a crucial instrument on the European Space Agency's (ESA) billion-euro JUpiter ICy moons Explorer. JUICE, as it is known, is due to launch on the last ever Ariane 5 rocket and it will be shaken to within an inch of its life as it leaves the Earth and embarks on its eight-year journey to Jupiter.

'You've spent ten years ensuring that this mission is going to be ready,' explains Dougherty, who also headed up the science definition team that proposed the mission. 'And then you watch it on top of a bloody huge rocket that you have no control over. Terrifying!'

"JUICE will be flying by Ganymede, Callisto and Europa to understand what lurks beneath their surfaces."

JUICE is ESA's upcoming foray into the depths of the Solar System, and it will focus on one of the major questions in planetary science and astrobiology: is there liquid water at Jupiter's moons? And then, the crucial follow-up question: if there is water, could it support life?

Life as we know it needs water. In the past, we thought that Earth was special because of its position in space. If our planet was closer to the Sun, the water would evaporate and, if it were further away, it would freeze. This location is known as the 'Goldilocks zone' because, just as the preferences of the light-fingered little girl in the fairy tale, it is neither too hot nor too cold.

Until recently, Mars had been the leading contender for finding life elsewhere in the Solar System (see page 58). Other scientists have pinned their hopes on planets outside our Solar System but similarly in a Goldilocks zone around a star. But now icy moons have also emerged as prime candidates for finding liquid water, especially after the discovery of geysers of water vapour on Jupiter's Europa, as well as on Enceladus, a moon of Saturn. So, JUICE will be flying by Ganymede, Callisto and Europa to understand what lurks beneath their surfaces.

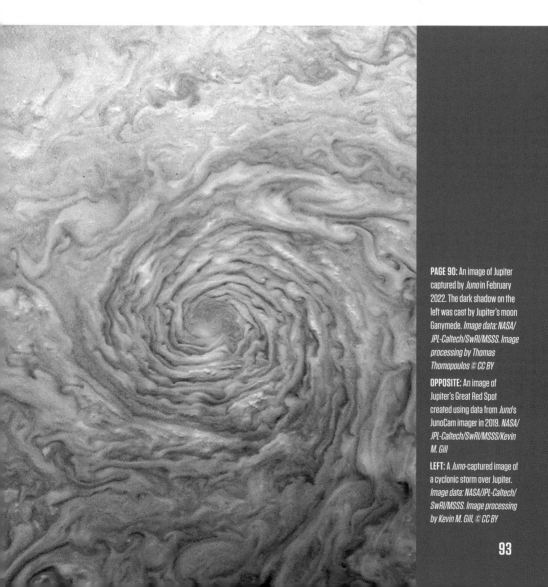

PAGE 90: An image of Jupiter captured by *Juno* in February 2022. The dark shadow on the left was cast by Jupiter's moon Ganymede. *Image data: NASA/ JPL-Caltech/SwRI/MSSS. Image processing by Thomas Thomopoulos* © CC BY

OPPOSITE: An image of Jupiter's Great Red Spot created using data from *Juno's* JunoCam imager in 2019. *NASA/ JPL-Caltech/SwRI/MSSS/Kevin M. Gill*

LEFT: A *Juno*-captured image of a cyclonic storm over Jupiter. *Image data: NASA/JPL-Caltech/ SwRI/MSSS. Image processing by Kevin M. Gill,* © CC BY

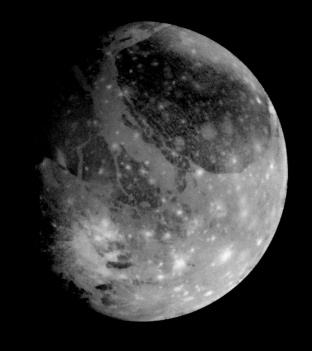

DEATH STAR

And each of these worlds is fascinating, mysterious and downright weird in its own right.

Ganymede, the Solar System's largest and most massive moon, disconcertingly looks like the Death Star from *Star Wars*, something it shares with Saturn's moon Mimas. Its mottled brown and white surface is covered with impact craters from other smaller (but speedy) objects hitting it. Ganymede is the only moon with its own internal magnetic field, which has scientists looking at this large pale moon with a great deal of interest.

Meanwhile, Callisto orbits Jupiter like a dirty disco ball, with thousands of white craters sparkling against a dull rainbow of colour. Its surface is the oldest and most densely cratered in the Solar System, and it could offer a unique perspective on how Jupiter's large moons formed billions of years ago.

TOP LEFT: A true-colour view of Ganymede, taken by the *Galileo* spacecraft in 1998. *NASA/JPL*

BOTTOM LEFT: The scarred surface of Jupiter's moon Callisto, in an image taken by *Galileo*. *NASA/JPL/DLR*

OPPOSITE: Jupiter's icy moon, Europa, again captured by *Galileo. NASA/JPL-Caltech/SETI Institute*

> **"Each of these worlds is fascinating, mysterious and downright weird in its own right."**

Europa, which is slightly smaller than our own moon, resembles a giant scratched marble, and has the smoothest solid surface in the Solar System. Scientists have confirmed the existence of water vapour on this moon and suspect a giant salty ocean lurks beneath an icy crust.

These moons are giving scientists ideas about other locations where they should be hunting for water. 'So now there's this vast region of space where you could potentially find life,' explains Dougherty.

JUICE, with its array of scientific instruments, is ready to investigate Jupiter and these three icy moons. But first it has to get there in one piece.

SPACE BUTTERFLY

The space probe will not have enough fuel to make it to Jupiter; it will need more energy. So, instead of a direct flight to the Solar System's largest planet, JUICE's eight-year journey will first involve flying by Earth, Venus and Earth, before a final pass around Earth again. It's a bit like swinging a rock at the end of a rope to make it go faster: JUICE will fly past planets in the solar neighbourhood, gaining energy, before its operators release the hypothetical rope and slingshot the spacecraft towards its final destination.

JUICE looks like a giant space-bound butterfly. It has a central, shielded and insulated trunk; numerous antennas sprouting from the body, as well as an almost 11m (36ft) boom; and two giant solar 'wings' unfurled on either side. These look like massive crosses, each about the size of the front of a bus. They will be vital for harnessing the Sun's power from Jupiter's distant neighbourhood, where solar energy is around twenty-five times weaker than on Earth.

'There are many individual technical requirements that we have faced before on previous missions,' says Cyril Cavel, project manager for Airbus, the company that built the spacecraft. 'What makes JUICE so special is that, for the first time, we accumulate all of these design challenges on one single mission.'

For example, the spacecraft will endure a wide range of temperatures, from the toasty environs of Venus – around 230°C (446°F) – to chilly Jupiter's –220°C (–364°F).

BELOW: An artist's impression of JUICE in its travels around the Jupiter system. © Airbus

OPPOSITE: Testing the solar array 'wings', June 2022. © Airbus 2022 JB Accariez – Master Films

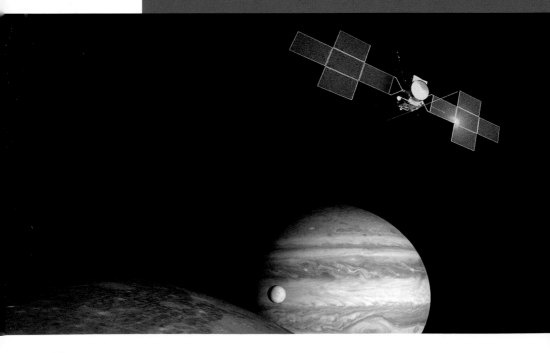

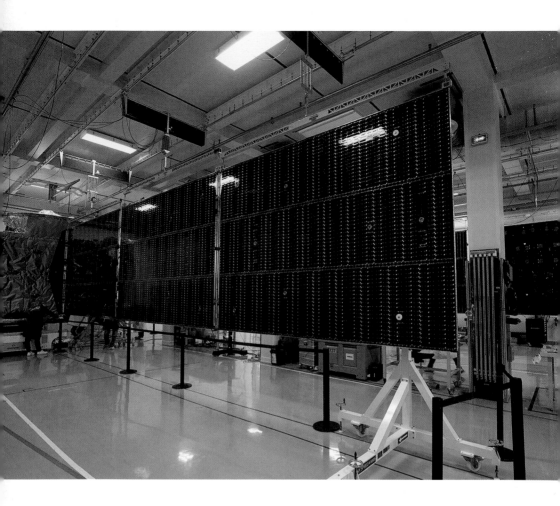

INTENSE RADIATION

Jupiter has the most intense radiation of any planet in the Solar System because of its super-strong magnetic field – and this radiation will eat away at all exposed external surfaces. Added to the environmental challenges, some of the signals that the instruments want to measure, such as the magnetic field within the oceans of Ganymede, are very weak. This means that the spacecraft's electric and magnetic fields must not interfere with the signals that the instruments are trying to detect.

To get around these difficulties, JUICE's body is swaddled in insulating material and shielded with lead foil, and the most sensitive instruments are stuck at the end of the spacecraft's boom – much like a microphone on a boom on a movie set. There are also many points at which the spacecraft will be out of communication with Earth, deaf and blind to scientists' instructions, so a lot of its functions – including self-diagnostics and problem-solving – are autonomous. In case something does go wrong, about half of the 6-tonne spacecraft's weight is taken up with spares.

'We have back-ups almost everywhere,' says Cavel, adding that even where there aren't back-ups, a plan can be made. 'There is only one engine, and even if it fails, we have other thrusters to complete the mission.'

LIQUID WATER

When it finally arrives at Jupiter in 2031, JUICE will be building on the legacy of past missions, beginning with NASA's *Pioneer* and *Voyager* probes in the 1970s. However, there have been only two previous dedicated missions to the gas giant: the 1989 *Galileo* probe and *Juno*, which arrived at Jupiter in 2016 and has been sending back spectacular images ever since. *Juno* is now on an extended mission until 2025, having completed its prime mission of thirty-five orbits of the planet

ESA's JUICE is fitted with ten instruments, some of which will be able to capture incredible images, from ultraviolet wavelengths to sub-millimetre wavelengths, as well as an optical camera that will show us what these surfaces look like to the human eye. The other instruments will collect as much information as the latest technology can offer: a radar to penetrate the crusts of three of its moons, and a spectrometer to hunt for chemical signatures and understand what the landscapes are made up of. Another tool

will decipher their atmospheres, and others will work to unpick their magnetic fields.

'We have selected the instruments to make sure that they work together, and we need all of them to answer the questions,' says Olivier Witasse, JUICE project scientist at ESA. 'We are going to be studying everything.' And he really does mean everything – how Jupiter and its moons interact, the composition of their crusts, their atmospheres, their magnetic fields and how these fields interact, among other things. However, there is one area of study that stands out. 'The most important part is to understand this liquid water. Is it 10 kilometres below the surface or 40, 50? We need to know to prepare for the next exploration.'

Jeffrey Kargel, a senior scientist at the Planetary Science Institute in Arizona, is excited to find out what the spectrometer will say about Europa's ocean, particularly the salts it contains – which could be the difference between a balmy ocean where life could possibly flourish and acidic icy water in which most life would struggle to survive.

In an arrangement as strange as Jupiter's – with its wobbly magnetic field, entourage of seventy-nine moons and the tantalising prospect of liquid water – scientists are particularly interested in data about the magnetic fields that permeate the Jovian system, in part because this can provide answers that other metrics can't.

'Magnetic fields describe the entire plasma environment,' explains Dougherty, who was principal investigator of the magnetometer on NASA's Cassini mission to Saturn. Plasma is a superheated gas, and makes up most of the Universe. 'But they also allow you to see inside planetary bodies.'

Her team is focusing on the magnetic field in Ganymede's ocean, but that is a big ask: they need to be able to detect the ocean's tiny magnetic field (about a nanotesla) against the background magnetic field of the rest of the moon, as well as Jupiter's

ever-changing field – which is more than a billion times stronger. 'When I lose sleep at night, it's about whether we're going to be able to resolve the ocean signature of Ganymede,' she says.

They will spend the next decade developing the computational models to separate the whisper of Ganymede's oceanic magnetic field from the background roar. But Dougherty will have to be patient, something which she admits does not come easily to her. JUICE's Ganymede gallivant is at the end of the mission – right before the spacecraft deliberately ploughs into the surface of that moon in 2035. That way, scientists will know where the craft crash-lands and make sure it doesn't smash into a planet where there may be life. As a bonus: you also learn a lot of things about a planet's surface when you hurl a heavy spacecraft at it.

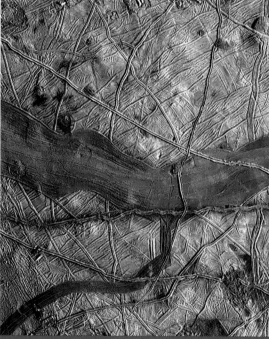

OPPOSITE: An artist's rendering of *Juno* in its elliptical orbit. *NASA/JPL-Caltech*
ABOVE LEFT: JUICE being worked on at the European Space Research and Technology Centre, in the Netherlands, in April 2021. *ESA-SJM Photography*
ABOVE RIGHT: The cracked surface of Europa – does an ocean lurk beneath its icy crust? Image captured by *Galileo*. *NASA/JPL-Caltech/SETI Institute*

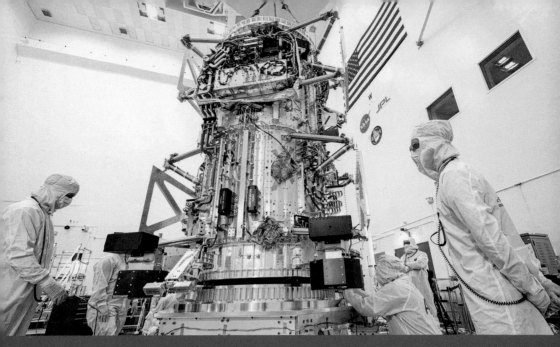

ABOVE: Engineers and technicians inspect the main body of NASA's Europa Clipper spacecraft. *NASA/JPL-Caltech/Johns Hopkins APL/Ed Whitman*
BELOW: The main body of the Europa Clipper spacecraft being transported to the NASA Jet Propulsion Laboratory in California. *NASA/JPL-Caltech/Johns Hopkins APL/Ed Whitman*
OPPOSITE: The Solar System's largest moon, Ganymede, captured alongside Jupiter by the *Cassini* spacecraft in 2000. *NASA/JPL/University of Arizona*

WILD AND COOL

Even as JUICE is poised to change our understanding of the Solar System and directly detect liquid water, planetary scientists and engineers already have their eyes on the next mission.

'In the same way that *Voyager 1* and *Voyager 2* pointed to *Galileo*, and *Galileo* pointed the way to JUICE,' says Kargel.

Hot on JUICE's heels, NASA's Europa Clipper mission is due for launch in 2024. 'We are very happy that there will be two spacecraft at the same time because, with two, you can do joint stuff and it's great,' says ESA's Witasse. The JUICE mission will only make two flybys of Europa, which will comprise about 5 per cent of its science, whereas Europa Clipper will focus solely on Europa and the possible water beneath its icy crust.

For Kargel, all the data from JUICE and then Europa will pave the way for the next step in planetary exploration. 'We'll be able to detect how thick the ocean is, how dense the ocean is, and how salty it is,' he says, speaking faster. 'We'll then be able to determine if we ever have any hope of sending a probe down into the ocean – because that would be the ultimate, wild, cool thing to do.'

As long as nothing goes terribly wrong during the wild ride of launch.

JUPITER AT A GLANCE

The first thing to know about Jupiter is that it is giant, really giant. It is twice as massive as all the other planets in the Solar System combined. It's also a gas giant, which is a large planet mostly made up of hydrogen and helium. Our Solar System has two gas giants – Jupiter and Saturn – and, unlike Earth, they do not have hard surfaces. They have solid cores surrounded by swirling gases.

Jupiter looks like it has cream, orange and brown stripes, which are caused by windy clouds of ammonia and water vapour that float in a firmament of hydrogen and helium. It has a giant red spot that is bigger than Earth but is actually a huge storm that has rampaged for centuries.

A swarm of moons surrounds Jupiter, almost eighty of them, and while Jupiter's tumultuous exterior cannot support life, some of its satellites may be able to. Like Saturn, Jupiter has rings, but they are much fainter than Saturn's and made of dust, not ice.

So far, nine spacecraft have visited Jupiter. Seven flew by and two have orbited the gas giant. NASA's Juno mission, the most recent, arrived at Jupiter in 2016.

There are a number of trips planned to our largest planet. As well as JUICE and NASA's Europa Clipper mission, China is also eyeing Jupiter and talking about launching a space probe. If JUICE finds proof of liquid water on Jupiter's icy moons, we may see even more visits to our Jovian neighbour.

ASK DR

What's at the centre of the Universe? What's inside Mars? How long would you survive in space without a suit? The Supermassive Podcast's resident astrophysicist, Dr Becky Smethurst, tackles your big questions on the nature of the cosmos.

BECKY

Q: We now have images of two different black holes. But we're told that light can't escape black holes, so how do we get pictures of them?

A: The Event Horizon Telescope has recently captured images of the supermassive black hole at the centre of the Messier 87 galaxy and our own galaxy, the Milky Way. Although we can't see the actual black holes, the telescope is so powerful that it's been able to detect the light coming from the material spiralling around them.

As hydrogen gas gets ever closer to a black hole, the extreme gravity accelerates it to tremendous speeds. This gives the gas molecules more energy so that they also become hot and start to glow. The bigger the

A dramatic view of the Milky Way over Fire Island, New York. *B.A.E. Inc./Alamy*

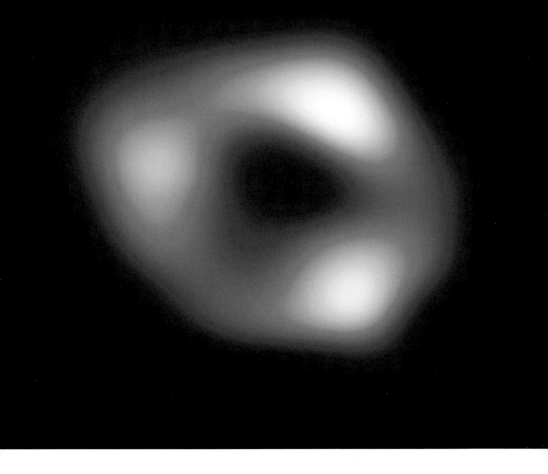

black hole and the more hydrogen gas there is, the brighter the regions surrounding the black hole will be. This light covers the entire electromagnetic spectrum, from X-rays and ultraviolet, to visible and infrared.

But the region in the very middle of this turbulent and hot material is the black hole itself, from which light cannot escape. So, while we see the ring of material spiralling around it, we also see the 'shadow' of the black hole in the middle. The black on the outside of the ring is empty space, whereas the black on the inside of the ring is where so much matter has been squished into such a small space that not even light can escape.

The Event Horizon Telescope is made up of radio telescopes around the world. Radio telescopes can be combined in such a way that they mimic a telescope that is the size of the Earth (see page 42). Previously, we only saw the centres of galaxies as fuzzy blobs but now we can resolve those blurs into rings. The bright blobs in the ring are called hotspots. This is where there is a slightly denser patch of material swirling around the black hole and therefore giving off more light.

Although, at first glance, these images just look like blurry, blobby doughnuts, they reveal so much about the nature of black holes and the strength and behaviour of gravity around them. They will allow us to do detailed tests on our theories of gravity to work out if we've missed something in the bigger picture of how the Universe works.

Q: How much time does it take for the James Webb Space Telescope (JWST) to collect enough data for a single image?

A: It depends on what JWST is looking at. Bright objects close by, such as the planets in our Solar System or nearby bright stars, won't need much exposure time before enough light is collected above the background noise to get a clear image. However, something fainter that is much further away – such as a distant galaxy billions of light-years from the telescope – will need a much longer exposure time.

To give you a rough idea how long JWST takes to collect enough light, let's compare it to its predecessor, the Hubble Space Telescope (HST). Back in 2003, Hubble took one of the most detailed images ever of a tiny patch of sky, which is now known as the Hubble Ultra Deep Field. It took Hubble a total combined time of just over 1 million seconds, or eleven and a half days, to collect enough visible light from the faintest and most distant objects. It then stared at the same patch of space detecting infrared light, which took a total observing time of 200,000 seconds, or around another two and a half days. JWST also detects infrared light and has some overlap in wavelength range with Hubble. However, because JWST has a much bigger mirror and can therefore collect more light, an exposure time of two and a half days on Hubble will take around three hours with Webb.

OPPOSITE: The stunning first image of the supermassive black hole at the centre of our Milky Way galaxy, Sagittarius A*, released in May 2022. *EHT Collaboration*

BELOW: A JWST alignment image released in May 2022. This shows part of the Large Magellanic Cloud, in a close-up from JWST's Mid-Infrared Instrument test image. *NASA/ESA/CSA/STScI*

"An exposure time of two and a half days on Hubble will take around three hours with Webb."

That's the good news. But getting time on telescopes is an incredibly competitive process. Most telescopes are oversubscribed by a factor of five to ten compared to how much time is available to use them.

When we astrophysicists put together an observing proposal to request to use a telescope, we employ what is known as an 'observing time calculator' tool. These calculators let you input the brightness of the object you intend to observe, along with its coordinates and a target 'signal to noise ratio' that you want to achieve. They then output an estimate for how long you will need to look at an object to get a decent image. With that information to hand, you can tot up how much time you will need on the telescope to do your science and hope your application is successful.

"The expansion of space is a never-ending battle between the expansion outwards and gravity pulling inwards."

Q: I'm confused about the idea of an expanding Universe. We are told that our neighbouring galaxy, Andromeda, is going to collide with the Milky Way. How is this possible if everything is expanding away from everything else?

A: Space is expanding, but objects like galaxies are not in fixed positions in space. The expansion of space is a never-ending battle between the expansion outwards and gravity pulling inwards. When objects are close together, gravity dominates over the expansion. This is why the Earth is not getting further away from the Sun as space expands; it's held in orbit by gravity as space expands around it. The same is true for the billions of stars, gas and dust held together in galaxies, such as in the Milky Way.

However, when objects are further apart from each other, such as the billions of light-years that separate galaxies, gravity is weaker and the expansion dominates, expanding the space between them and causing them to get further apart. The majority of galaxies we see in the sky all appear to be moving away from us due to the expansion of space between us and them. The Andromeda galaxy, however, bucks that trend.

Andromeda is our closest massive neighbour galaxy at roughly 2.5 million light-years away, and it contains over a trillion stars – many more than the Milky Way's roughly 100 billion stars. The gravitational attraction between the Milky Way and Andromeda is strong enough to pull them closer together despite the expansion of space between them. It means that in roughly 2–4 billion years' time, the two will merge together to become a much larger blob of a galaxy, dubbed Milkomeda.

In this forthcoming merger, it's very unlikely that any two stars will collide, since there's just so much space between them. The nearest star to the Sun is four light-years away, or 38 trillion kilometres (23.6 trillion miles), whereas the stars themselves are only between 1 and 10 million kilometres (0.6–6.2 million miles) wide. You could fit 27 million Suns end to end in the distance between it and its nearest star. So, the odds of any two stars coming close enough to collide is vanishingly small just because of the huge scales involved.

OPPOSITE: The Hubble Ultra Deep Field – a view of nearly 10,000 galaxies taken in 2003. The smallest, reddest galaxies may be among the most distant known, existing when the Universe was just 800 million years old. *NASA, ESA, and S. Beckwith (STScI) and the HUDF Team*

LEFT: A bird's-eye view of part of the Andromeda galaxy taken by Hubble in 2017. It is the sharpest image ever taken of our galactic next-door neighbour. *NASA, ESA, J. Dalcanton, B.F. Williams, and L.C. Johnson (University of Washington), the PHAT team, and R. Gendler*

Q: Have we found any exoplanets with moons, and are they comparable to the size that the Moon is to Earth?

A: So far, we only have twenty candidate exomoons. These are moons in orbit around planets beyond our Solar System. However, we only have a direct image of one of them. Scientists have proposed the existence of the others to explain data collected by telescopes, such as the Kepler space telescope, observing the brightness of stars.

If an exoplanet passes in front of a star during its orbit, the brightness of the star will dip by a tiny fraction, which we can then detect. If there is also a moon in orbit around that exoplanet then it will cause a

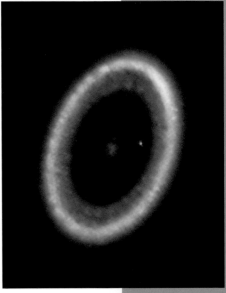

secondary dip in the brightness. From how much the brightness of the star dips for the pass of each object, we get an estimate of how big the exoplanet and exomoon are. But even if we detect these secondary dips in the brightness of the star, there could be other explanations for that change.

In 2021, astronomers working with the ALMA array of radio telescopes claimed to have taken an image of an exomoon forming around the exoplanet PDS 70c. Although it just looks like a fuzzy blob, the astronomers managed to model the light detected and suggested there was a ring of debris around the exoplanet that could form into an exomoon.

If you're looking for a planet–moon system most like our own, you are going to have to be patient for a while longer yet. With only a handful of candidate exomoons, it's difficult to make any concrete claims for their size relative to their planets. Our Moon is around 1.2 per cent the mass of Earth,

and there have been a few exoplanets with masses larger than Jupiter that have candidate exomoons with a similar mass ratio.

Astronomers have, for example, suggested that WASP-12b, which is an exoplanet roughly one and a half times the mass of Jupiter but which orbits very close to its star, has an exomoon that could be up to 6.4 times the mass of the Earth, putting the mass ratio of 1.4 per cent at a similar value to the Earth–Moon system.

The idea of exomoons around massive gas giant exoplanets close to their stars is an intriguing one for the existence of life beyond Earth. If the exomoon is rocky and found in the habitable (or Goldilocks) zone around the star (not too hot and not too cold), then the conditions for life could be just right. Given how many 'hot Jupiters' we have detected (Jupiter-sized planets orbiting closer to their stars), it could be that the majority of life in the Universe is found on moons, rather than planets.

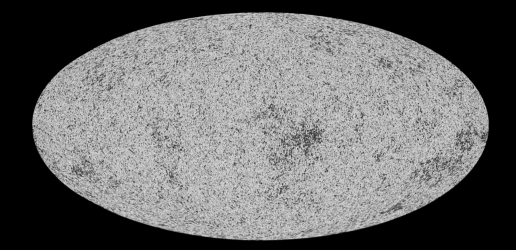

Q: If we can't see the centre or the edge of the Universe, then how can we possibly guess its lifespan?

A: So, first and foremost, there is no centre of the Universe. The Big Bang happened everywhere, all at once. Expansion is happening everywhere as well. Although it appears that, as we look out, we see everything moving away from us, that's just our perspective. It's nothing to do with the Universe itself. If you hopped to another galaxy, you'd see exactly the same thing. We are each at the centre of our own observable Universe because we can see the same distance in all directions around us. So, essentially, you are the centre of the observable Universe!

We don't need to see the 'edge' of the Universe either (if there even is one!) to work out how long it's been around for or what's going to happen in the future. The greater the distance you look in the Universe, the more expansion you see. So, with every extra million kilometres you step, galaxies appear to be moving away faster than those closer by, because there's more space that's expanding between us and them. Therefore, when we measure the expansion, we're measuring how much the speed changes (in, say, kilometres per second) for every kilometre you go. Since we know how big the observable Universe is from side to side (from how far we can see into space), we can work out how long it must have been expanding for, at the rate that we measure. This gives us an estimate for the Universe's age.

Q: What would happen to your body in space without a working suit or suitable protection?

A: A lot of people think that the big issue in space is the temperature. Space is incredibly, incredibly cold. While it's true that the average temperature of space is around 2.7 degrees above absolute zero (–273°C/–459°F), if you went out into space with no suit, you would actually feel warm if you were in the sunshine.

The reason for this is that space is a vacuum. There are no air molecules – and air molecules are essential for heat transfer. In air, molecules collide all the time and they gain and lose energy – that's how heat is transferred to and from you in Earth's atmosphere. But because space is a vacuum, that doesn't happen. The only way for energy to be lost in space is by infrared radiation and it would take a very long time for heat to radiate away from your body and to freeze in space.

So, if it's not the cold that will kill you, what will? Well, it's that lack of air in the vacuum again. If you were being flung out from your craft's airlock into space, your instinct might be to take a big deep breath, but as the pressure drops, the air in your lungs would compensate for that pressure drop by expanding to bursting point. Scuba divers will be familiar with this concept and it's the reason why you should never hold your breath underwater while scuba diving in case you change depth (and therefore water pressure) very quickly.

The whole 'no air' thing isn't ideal but the main problem of being in a vacuum is that the boiling temperature of water is lowered. So, all of the water in your body will start to evaporate. You'll get incredibly dry eyes and an incredibly dry mouth, but also the water in your bloodstream will start to boil. With nowhere to go, it will get trapped under your skin so that it looks like your skin is bubbling. Along with the water, the oxygen in your blood will also evaporate and your blood will become deoxygenated in about fifteen seconds. When deoxygenated blood hits your brain, you will pass out. So, thankfully, you won't experience any of these unpleasant sensations, but your fellow astronauts will only have a minute or so to save you.

OPPOSITE: This map of the Universe was produced by ESA's Planck mission. It shows the oldest light in our Universe, called the cosmic microwave background. This was imprinted on the sky when the Universe was 370,000 years old. The colours represent tiny temperature fluctuations that correspond to regions of slightly different densities , which eventually evolve into the galaxies of today. *ESA and the Planck Collaboration*

LEFT: Astronaut Carl J. Meade performs an untethered spacewalk. *NASA*

"If we do ever find life on Mars, the big question will be whether it is at all similar to the life we see on Earth."

Q: What is the current model of the internal structure of Mars and how does that compare to Earth?

A: Thanks to NASA's *InSight* lander, which touched down on Mars in November 2018, we now have a much better idea of the interior structure of the planet. *InSight* is fitted with a seismometer to detect marsquakes – equivalent to our own earthquakes.

Geologists use earthquakes to determine the internal structure of our planet by detecting how they travelled through the various layers of solids and liquids under the crust. Starting at the surface and travelling inward, we know that the Earth has a rocky crust which, on average, is between 35 and 200km deep (22–124 miles), and has a liquid mantle of magma, surrounding a fluid outer core made mostly of iron and nickel. Finally, in the middle, Earth has a solid iron and nickel core which is around 20 per cent of the total radius of the planet.

Mars has its similarities. The latest data from the *InSight* lander suggests its crust is anywhere from 20 to 40km (12–25 miles) thick, again with a liquid mantle below it, surrounding a liquid iron core. Mars's core, however, makes up just over half of the entire planet, which is very different to our planet.

An artist's impression of the *InSight* lander on Mars, showing its instruments deployed on the Martian surface.
NASA/JPL-Caltech

Even marsquakes are generated differently to earthquakes. On Earth, tectonic plates move and slip past each other to create an earthquake. However, Mars is not as geologically active and so does not have tectonic plates. Instead, its rocky crust has been cooling and therefore shrinking since it formed billions of years ago. As it shrinks, if there are any faults in the rock, the pressure releases and a marsquake is generated.

It's intriguing to think about how the different interior structures of Earth and Mars led to their conditions today. The liquid metal in Earth's core is key to having a protective shield, our magnetic field. It's so hot that the electrons and protons in atoms are no longer bound together. As heat is transferred around by convection, these charged particles move in patterns that generate the Earth's magnetic field, which protects us from high-energy charged particles from the Sun. These get funnelled to the poles and interact with elements in the atmosphere to produce the aurora.

We think that, once upon a time, Mars's atmosphere was thick and lush, providing the perfect conditions for life. However, Mars doesn't have a very strong magnetic field, suggesting its liquid core does not have the same movement of charged particles as Earth's. Without a strong magnetic field, Mars has been bombarded with solar radiation that has slowly stripped the Red Planet of its atmosphere, leaving a planet that now appears to be a barren wasteland. It's why the *Perseverance* rover is currently on the Martian surface hunting for fossilised life from a time when the conditions on Mars may have been more favourable.

If we do ever find life on Mars, the big question will be whether it is at all similar to the life we see on Earth: does it have a common origin? Did comets and asteroids perhaps bring the building blocks for life as they collided with each of the planets? Or did those same impacts throw material off Earth into space that made its way to Mars to seed life, or maybe the other way round? By investigating Mars, we might just be able to answer one of the biggest questions humanity has ever asked: where did we all come from?

Q: When we look at the oldest galaxies we can see, is there reason to believe they formed or behaved any differently from the newer ones?

A: Astronomy is like a time machine. If you look at things a long way away, then – because it takes photons of light from those objects so long to reach us – you are seeing them as they were a long time ago. And in the case of the most distant galaxies, that's over 13 billion years ago! We are seeing what the ancestors of galaxies close to us today will have looked like. They haven't yet settled into the kind of shapes that we see now. They don't have either the characteristic spiral or elliptical shape that we associate with nearby galaxies and instead tend to be a lot smaller.

The big galaxies we see around us are ones that have merged and collided with neighbours to develop their shape. Early galaxies are different, but that's because they have not yet evolved. There aren't galaxies like this around in the same way today. This is also a way of proving the Big Bang theory is correct. It shows that the Universe was definitely very different in the past because the galaxies looked so different.

Q: Do all galaxies spin in the same direction? If not, what determines the direction of the spin?

A: When we look at galaxies, about half of them appear to be turning in one direction, clockwise, with the other half turning anticlockwise. And what that implies is that there isn't a preferential direction in which galaxies rotate. So, the roughly 50:50 split comes down to our point of view. For example, if you look at a rotating wheel or a fidget spinner from one side, it'll look like it's moving clockwise. And if you look at it from the other side, it will be anticlockwise. It's the same thing for the uncountable number of galaxies we see spread out across the Universe. It would be very, very weird indeed if that wasn't the case; it would suggest there was some law of physics that we don't yet know about that could cause galaxies to form rotating preferentially in a given direction.

And as for how they actually rotate, it's different for spirals and ellipticals. In elliptical galaxies, all the stars are orbiting the centre but in different directions and planes, resulting in something roughly spherical in shape (hence elliptical). I often describe the stars in these galaxies as behaving like bees in a beehive. As for spiral galaxies, usually all the stars are orbiting the centre in the same direction in a single flat plane. But that's not always the case; if there's been a recent collision or merger of two galaxies then the rotation can be altered (this is how we think the beehive-like structure of ellipticals is eventually formed). For example, you can have the stars in the centre moving in the opposite direction to the ones on the outskirts; this is called a counter-rotating disk and can cause chaos!

Q: Do we know if the Moon ever had a magnetic field?

A: That's a great question because if the Moon had a molten core, it must have had moving electric charges, which produce a magnetic field. And, yes, there's evidence that the Moon could actually have had a magnetic field at some point in the past.

We know it doesn't any longer, presumably because the core has solidified and those charges have stopped moving around. What's really cool is that some of the evidence that we have for the Moon having had a magnetic field in the past comes from the lunar rocks that were brought back to Earth by the Apollo missions. You can see the alignment of metals in the rocks that suggests there was a lunar magnetic field about 4 billion years or so ago, right after the Moon formed. But the big debate that's still ongoing is whether the Moon had a global magnetic field or whether there were little bursts of magnetic fields being created from impacts during periods of heavy bombardment in the early days of the Solar System. Those collisions could have generated the magnetic fields that would cause this alignment of metal atoms in the Apollo Moon rocks.

To find out which it is, we will need more rocks collected from different locations on the Moon and also rocks of different ages. That's one scientific reason to go back to the Moon.

OPPOSITE: Hubble captured this image of a group of interacting galaxies some 400 million light-years from Earth. *NASA, ESA and M. Livio, and the Hubble Heritage Team (STScI/AURA)*

BELOW: Apollo 17 astronaut Harrison Schmitt was the only geologist to walk on the Moon. *NASA Johnson*

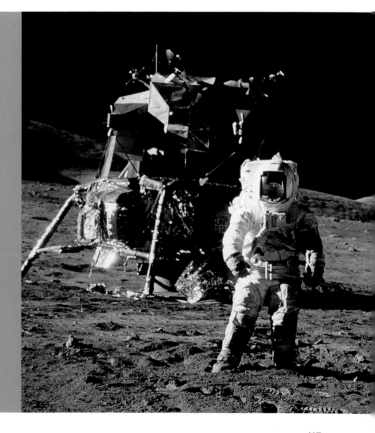

"You can see the alignment of metals in the rocks that suggests there was a lunar magnetic field."

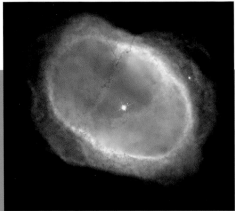
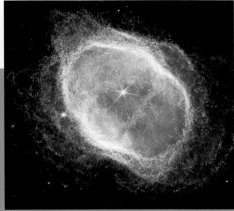

"Although the images are going to be very different, I think they'll be just as visually stunning as those of the Hubble Space Telescope we've become familiar with."

Q: How do images from the James Webb Space Telescope compare to those from Hubble which have inspired the world for decades?

A: JWST observes in a different wavelength to the Hubble Space Telescope – it sees in infrared light, not visible light. That means we're going to be able to see different things. Infrared light is a longer wavelength, so you can see through gas and dust, the things that often give rise to those glorious nebula pictures we see from Hubble that everyone knows and loves. So, while at first it might seem like the images we'll get from JWST will be less beautiful without all of this gas and dust that blocks the light and provides contrast, we'll still be able to see things that we've never been able to before in such detail.

But JWST also has a bigger telescope. It has a 6.5m (21ft) wide mirror to collect light compared to the Hubble's 2.4m (7.8ft) wide mirror. This is important, because the bigger the telescope, the smaller the thing you can pick out in the sky; the resolution gets better. So we're going to see much more detail in these images than we've seen before with Hubble. This greater detail is definitely going to give us some incredible images, especially of objects in the Milky Way for exoplanet science or perhaps studying star formation in gas clouds. It's really going to be a game-changer for a lot of different fields. Although the images are going to be very different, I think they'll be just as visually stunning as those of the Hubble Space Telescope we've become familiar with, just for different reasons.

Q: Why do all the gas giants have rocky moons when they were formed in an area of space that led the planets to being gas giants?

A: The moons are essentially the leftovers that formed after the planets did. Gas giants were formed in a region of the Solar System where you could have hydrogen that wasn't affected by the solar wind, but you still had all these clumps of rock as well. We call them planetesimals – that is, baby planets. Multiple collisions between these smaller bodies meant they came together and all of them eventually got rounded up over time.

If we take Jupiter as an example, it will have been a big messy clump – with lots of bits of rock embedded in hydrogen. The hydrogen was all attracted to the biggest object in the centre and this ended up being Jupiter, which has a dense solid core.

Once that settled, everything that was left over – the rocky lumps – then formed the moons. Jupiter, Saturn, Uranus, Neptune, they're so big that if an asteroid wanders too close, it's going to get captured by the planets' gravity. Jupiter has been collecting lumps of rock that stray too close for around 4 billion years and even has two clumps of asteroids that trail and lead Jupiter in its orbit that we call the Trojan asteroids.

OPPOSITE: The Southern Ring Nebula, as captured by Hubble (left) and JWST (right). *Hubble Heritage Team (STScI/AURA/NASA/ESA) / NASA, ESA, CSA, and STScI*

BELOW: Jupiter and its icy moon Europa, captured by Hubble in August 2020. *NASA, ESA, A. Simon (Goddard Space Flight Center), and M. H. Wong (University of California, Berkeley) and the OPAL team*

ARTEMIS: BACK TO THE MOON

As humanity prepares to return to the Moon, Sue Nelson examines the ambitious Artemis programme and the important milestones that need to be reached before 'one small step' becomes a reality for both men and women for the first time in history.

Thousands of years before Katniss Everdeen, there was Artemis. A Greek goddess of hunting and wild animals, she is often depicted in statues with a bow of arrows slung across her shoulders and a deer at her feet.

At first glance, it seems a strange choice of name for NASA's programme to return to the Moon. However, in Greek mythology, Artemis had several titles. Daughter of Zeus and sister of the Sun god Apollo, Artemis was also the Moon goddess and a protector of women. How apt, then, that the Artemis missions will include placing the first woman upon the Moon.

It has been over fifty years since the last human walked on the lunar surface. The final footprints belong to Apollo 17 astronaut Eugene Cernan (see page 28), one of only twelve people – all American, all male – to have achieved this honour. Things will be different this time. Unlike the Apollo missions, Artemis is an international project that involves space agencies and industrial partners across the world. Although the first Artemis crew candidates are all American, astronauts from the European Space Agency (ESA), the Canadian Space Agency (CSA) and other nations will join this exclusive list in the future.

WHY BOTHER?

Since many believe we've already 'been there, done that' when it comes to the Moon, some people have asked: why bother?

Without wishing to repeat the words of *Star Trek*'s James T. Kirk, the primary reasons are science and – yes – exploration. There remains much to learn about our nearest astronomical neighbour in terms of how it formed and evolved over the last 4.5 billion years.

The Moon, for instance, has around the same surface area as Africa. Imagine basing your entire knowledge of the continent on only six points of contact.

Exploring around the landing site during the Apollo missions of 1969–72 was also limited in case the rover broke down. With only a finite air supply, the astronauts would have had to walk back to the lunar module, so they could not stray far. There's a lot more to explore.

Since water ice was confirmed on the Moon in 2008 by a NASA instrument on board India's Chandrayaan-1 spacecraft, the likelihood of building a lunar base has also increased. SpaceX CEO Elon Musk is particularly keen to build one and has already named his proposed habitation Moonbase Alpha, after the Moon base in Gerry Anderson's 1970s TV sci-fi series, *Space: 1999*.

Now we know that there are lunar resources for supporting life and, potentially through splitting water into its component parts, hydrogen fuel, this lays the groundwork for a crewed mission to Mars.

But that's a few decades ahead. Stage one is returning humans to the lunar surface, kick-starting a renewed era of human space exploration. There is a huge ongoing global collaboration for the Artemis programme. But there are several challenges that need to be overcome to make this ambition a reality.

THE SLS ROCKET

The first question is how to get there. A Saturn V rocket first propelled the Apollo missions to the Moon in 1969. Not surprisingly, over half a century later, the Artemis programme required a launcher with an

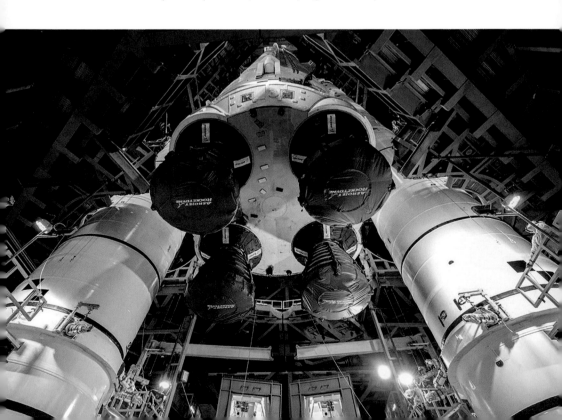

upgrade and design changes. The result is the Space Launch System, or SLS, rocket.

The SLS, at 98m (321ft) high, is only slightly taller than a Saturn V, but, in terms of thrust at launch, it is 15 per cent more powerful. Like the Saturn V, the SLS is also split into stages.

At the bottom of the rocket are four RS-25 liquid rocket engines, of the same type that powered the Space Shuttle. At the sides are two solid rocket boosters attached to a distinctive orange-coloured core stage, which contains the propellants: the liquid hydrogen fuel and the liquid oxygen that is used as an oxidiser to make the fuel burn.

Above the core stage is a 9m (29ft) tall conical launch vehicle stage adapter. Rather like that additional travel plug you need

when going abroad, this simply connects the core to the interim cryogenic propulsion stage. Once at low Earth orbit, this stage propels the next and final part towards the Moon: the Orion spacecraft and the launch abort system at the top.

THE ORION CAPSULE

The Orion spacecraft consists of two parts: a crew module or capsule and a service module. At first glance, the Orion and Apollo capsules look remarkably similar – a sort of metal shuttlecock-shaped craft with the rounded rubber end removed. But we have entered the digital age since Apollo's demise and so inside there are fewer manual switches and, instead, three

RIGHT: Artemis II's heat shield being constructed in July 2020 – essential for protecting the astronauts and capsule during reentry through Earth's atmosphere. *NASA/Isaac Watson*

OPPOSITE: An ESA cross-section of Orion's service module © *ESA– K.Oldenburg*

large digital display screens. The capsule is also bigger, which allows for a crew of four instead of three.

It also has the largest heat shield ever designed for a spacecraft carrying astronauts. This heat shield will protect astronauts if there is a solar particle event (SPE), when the Sun belches out charged particles that can disrupt communications on Earth and damage an astronaut's body tissue in space.

If an SPE is detected, the astronauts will shelter beneath the capsule floor, protected by the heat shield on one side and a 'pillow fort' of stowage bags on the other. At 5m (16.4ft) across, it's still not a huge amount of living space but, once Orion reaches orbit, the astronauts can fold and stow away seats for more room.

The Orion capsule's maiden test flight took place on 5 December 2014 from Cape Canaveral, Florida. I was there to watch and experienced the stomach-resonating, crackling roar of lift-off first-hand. Trust me, it's loud. The capsule's orbit system was then successfully tested in 2019 so, as far as Artemis is concerned, it is a go for launch.

THE EUROPEAN SERVICE MODULE

The other part of the Orion spacecraft is the crucial section beneath the capsule: the European Service Module (ESM), made by ESA. Its role is to provide the capsule with power, propulsion and life support. Without its contribution, Orion couldn't reach the Moon.

ESA has a good track record of working with NASA and the ESM evolved from ESA's Automated Transfer Vehicle, which delivered supplies to the International Space Station (ISS) and helped maintain it in orbit by boosting its engines to raise the station's altitude.

A circular spacecraft 4.5m (14.8ft) in diameter, the ESM has four 7m (23ft) long solar panels. Thirty-three thrusters using three types of engine allow the manoeuvrability and position control required for Orion to be aligned towards its destination. Over 11km (6.8 miles) of cables form the avionics that will send commands and receive information from Orion's sensors.

The ESM also contains all life support for the astronauts in the form of onboard oxygen, nitrogen, food and water. In addition, radiators and heat exchangers will ensure everything inside is kept at a comfortable temperature. Considering the average temperature in space is –270°C (–458°F), this is especially important for the astronauts.

Incidentally, the module is covered in Kevlar, the material used for bulletproof vests. This will reduce any damage from micrometeorites and help further protect the crew inside. This same material comes into play when Orion makes its return to Earth via a splashdown into the Pacific Ocean. Eleven parachutes, made from a Kevlar–nylon hybrid, will slow the spacecraft down so that it won't do the equivalent of a painful belly flop. Finally, an amphibious transport ship will be waiting nearby to pick them up and bring the astronauts and the capsule back to dry land.

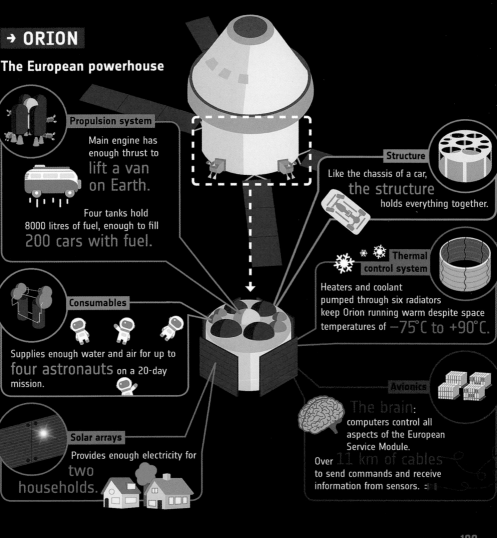

→ ORION

The European powerhouse

Propulsion system
Main engine has enough thrust to
lift a van on Earth.

Four tanks hold 8000 litres of fuel, enough to fill 200 cars with fuel.

Structure
Like the chassis of a car, the structure holds everything together.

Thermal control system
Heaters and coolant pumped through six radiators keep Orion running warm despite space temperatures of $-75°C$ to $+90°C$.

Consumables
Supplies enough water and air for up to four astronauts on a 20-day mission.

Avionics
The brain: computers control all aspects of the European Service Module. Over 11 km of cables to send commands and receive information from sensors.

Solar arrays
Provides enough electricity for two households.

THE ARTEMIS ASTRONAUTS

So who are the lucky people who will be making history on this new, cutting-edge mission? NASA has selected eighteen astronauts but, as yet, we don't know who will be chosen for each mission.

What we do know is that the Artemis team is racially diverse, with an equal male–female split and an impressive range of skills and experience. For example, Stephanie Wilson, who was originally selected in 1996, is the earliest qualified astronaut in the team and has three Shuttle flights under her belt.

Nicole Mann and Christina Koch were both selected for astronaut training in 2013. Koch currently holds the record for the longest single spaceflight for a woman, while Mann, a former US Marine Corps second lieutenant with a master's in mechanical engineering, has flown forty-seven combat missions in Iraq and Afghanistan but is yet to earn her space wings.

Jonny Kim is another multi-talented member of the team. A medical doctor, former Navy SEAL with a second degree in mathematics, he completed his two-year training in 2019 and is one of eight members selected from the class of 2017 who, along with Mann, is yet to fly.

Captain Scott Tingle, a former US Navy pilot, spent 168 days as flight engineer onboard the ISS from December 2017 until June 2018. Together with a colleague, his spacewalk installed a mechanical gripper on the station's robotic arm.

All the other astronauts are equally inspirational. So, whoever gets to stand on that lunar surface next will be an outstanding representative of humanity. But before that new moment of space history happens, let's look at the missions leading up to it.

ARTEMIS I

This first Artemis mission is the foundation of what will follow in terms of returning humans to the Moon. It is a test flight carrying engineering equipment and instruments to monitor the performance of both Orion and the SLS. Although uncrewed, there will be a 'Moonikin' on board: an astronaut mannequin carrying sensors to assess the acceleration, radiation and vibration for the (real) men and women to follow.

This test mission is crucial – but the launch itself has been far from smooth. President J. F. Kennedy once said that we choose to embrace challenges such as going to the Moon 'not because they are easy, but because they are hard'. Artemis is no different. The first launch attempts in September 2022 were scrubbed, primarily due to liquid hydrogen leaks during fuelling. At the time of going to press, Artemis I was still sitting on the launchpad. It is a nail-biting period for the engineers, as they work to fix problems while the world watches and waits.

So what *should* happen once that huge rocket finally gets off the ground? Two minutes into the mission, after consuming their propellant, the boosters will jettison, shortly followed by the service module fairings and launch abort system. After the core stage separates, Orion will orbit the Earth with the ESM and cryogenic propulsion stage (CPS) attached. During the first orbit, the ESM will extend its solar arrays to produce its own power and will use the thrust from the CPS to increase its speed for the twenty-minute translunar injection manoeuvre. It will also release ten small science satellites called CubeSats. The spacecraft will then separate from the CPS and head towards the Moon, with the vital ESM providing propulsion and electrical power for the four-day journey.

On day six, Orion will fly approximately 100km (62 miles) above the Moon at its closest approach and extend its orbit to 64,000km (40,000 miles) away from the Moon. Unlike Apollo, it will use a deep retrograde orbit – moving in the opposite direction from the Moon orbiting the Earth. On its return home, once the ESM separates, Orion will splash down in the Pacific Ocean.

Incidentally, there will also be a flash drive on board containing over 1.7 million names after a social media campaign. Yes, of course, mine is on it. More importantly, the heat shields will be tested on returning to Earth as they need to withstand temperatures of up to 2,760°C (5,000°F). Since the next mission will carry astronauts, this is one test that can't fail.

LEFT: Technicians rehearse booster-stacking operations at NASA's Kennedy Space Center in Florida in September 2020, ahead of the Artemis I launch. *NASA/ Kim Shiflett*

ARTEMIS II

The second mission ramps up the excitement levels. This is the one that will take four astronauts – three Americans and a Canadian – on a ten-day test flight around the Moon, which will set a record for the furthest distance humans will have travelled from the Earth. The orbit will take them 7,400km (4,600 miles) beyond the far side of the Moon.

Before they leave Earth's orbit, the crew will ensure the life support system is performing well and, during its journey to the Moon, the mission will check that deep space communications and navigation systems are all ready for Artemis III.

ARTEMIS III

This is the biggie. The Artemis III mission, with another crew of four astronauts, will see both a man and woman walk together on the lunar surface for the first time.

In 2020, SpaceX, Blue Origin and Dynetics were in competition to develop a human landing system (HLS) for this mission. The idea is that it will dock with Orion or a future orbiting staging post called the Gateway (see opposite). NASA selected SpaceX as the sole contractor to go forward in April 2021, with its massive reusable rocket-spacecraft-lander combo called Starship, much to the other companies' chagrin. Blue Origin even filed a lawsuit, which held up SpaceX's development work.

In October 2021, the US Senate Appropriations Committee told NASA it had to choose a second company to develop a Moon lander and ensure competition. So other companies will potentially build new landers to the Moon beyond Artemis III, but for now SpaceX remains the company that will produce the lander to return human beings to the Moon for the first time in over five decades.

The landing site is yet to be confirmed but is likely to be somewhere around the Moon's south pole.

RIGHT: The fourth ESM for Artemis IV, seen here in Turin, Italy, before being transported to Bremen, Germany, in June 2022. Sixteen companies in ten countries are supplying the components for the Artemis missions. © *Thales Alenia Space*

OPPOSITE: The Orion capsule approaches the Gateway staging post in this artist's impression. *NASA/Alberto Bertolin*

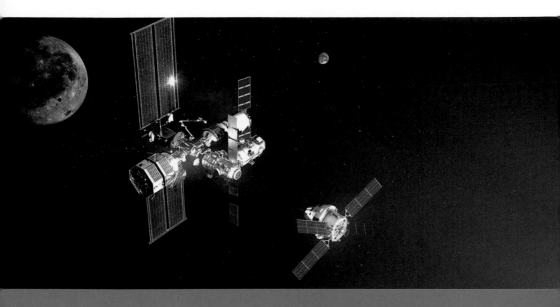

THE GATEWAY TO THE FUTURE

Once the Artemis III mission has been completed, future travel to the lunar surface will eventually see astronauts and uncrewed missions use an orbiting space station around the Moon as a staging post for further exploration and science. This lunar equivalent of a motorway service station is called the Gateway.

The Gateway will be a thousand times further out in space than the ISS and will act as a long-term base for astronauts to visit the lunar surface. It will be built over several stages, starting with two modules: the PPE (Power and Propulsion Element) and HALO (Habitation and Logistics Outpost). These modules will be integrated before launching to the required lunar orbit.

The European Radiation Stations Array (ERSA) will monitor the radiation environment to ensure the astronauts are kept safe and the Canadian Space Agency, which made the ISS's robotic arm, is providing advanced robotics.

ESA is providing the Gateway's International Habitat (I-Hab) – where the astronauts will sleep – and the ESPRIT module (European System Providing Refuelling, Infrastructure and Telecommunications). Yes, space scientists love acronyms. ESPRIT will deliver additional advanced lunar communications, a science airlock to deploy experiments and CubeSats (miniature satellites that are used for scientific research), as well as refuelling for the Gateway. It's likely that ESPRIT will be the most popular module with astronauts, however, because it will have a cupola like the one on the ISS. Imagine those views of the Moon from the window. The Japan Aerospace Exploration Agency is also contributing towards habitation components and logistics resupply.

Artemis IV onwards will start building the Gateway and eventually the outpost will support not just a new era of lunar discovery and permanent human habitation, but also act as a staging post for missions to Mars. It may have taken us over half a century to get back to the Moon, but the Artemis programme will inspire a new generation now that lunar travel, a working Moon base and, in due course, footsteps on Mars become a reality.

LOOK UP:
WHAT TO SEE IN THE NIGHT SKY IN 2023

The Supermassive Podcast's other resident astronomer, Deputy Director of the Royal Astronomical Society Dr Robert Massey, suggests what to look out for in the year ahead.

The year 2023 is packed full of exciting astronomical events that involve the planets, meteor showers and eclipses of the Sun and Moon. You don't need an expensive telescope to see them. This short, month-by-month guide is mostly for the northern hemisphere, but some events, like the eclipse in April, are highlights of the south. Southern skies are also incredible, with a great view of the brightest part of the Milky Way and this year planets like Saturn are much higher in the sky.

Wherever you are, we want to explain what you can see and how to find some of the best objects in the sky, just using your eyes, or maybe that previously unloved pair of binoculars or small telescope you might have lying around at home.

It's always better to try to escape the lights of towns and cities for the magic of a truly dark sky. But even in urban areas you can still see and easily learn the constellations and the names of their stars, find the Moon and the brightest planets. If you are lucky enough to be away from significant light pollution, then you can properly take in the Milky Way and find fainter objects like the galaxies and the beautiful clouds of gas and dust called nebulae – beloved targets of astronomy photographers.

Finally, do remember how quickly you can feel cold when standing still looking up at the stars. Even after a long, hot day in summer, the nights can be chilly. Warm clothing, probably more than you think you need, is a must, particularly if you plan to set up a telescope or take pictures of the sky.

JANUARY

The constellation of Orion, named for the mythical Greek hunter, takes centre stage in the January sky. Orion's belt of three stars and the box of four stars around it appear high in the south as it gets dark. Like all constellations, Orion looks this way from our Solar System but if we could travel across the galaxy it would be unrecognisable as its stars are at very different distances.

Orion's brightest star is Rigel, at the bottom right of the constellation – a blue supergiant star 850 light-years away, cal-culated to be more than 120,000 times as luminous and eighty times as big as the Sun. Also a supergiant, but this time red, at the top left is Betelgeuse, 550 light-years distant.

Hanging down from the belt is the fainter sword of Orion, which has a misty patch in the middle. This is the Orion nebula, one of the brightest and most famous 'stellar nurseries' packed with newborn bright stars. The nebula and the stars it contains are easy to view with a pair of binoculars or a small telescope, and are always a favourite target for photographers and amateur astronomers alike.

Much closer to home, four planets are easy to find this month. Jupiter, brighter than any star, stands out in the southwest after sunset. With a pair of binoculars or a telescope you can see its four largest moons and the shape of the giant planet, which rotates so quickly that its atmosphere bulges out at the equator. Above Orion in Taurus, the bull, is Mars, just past its best, but still a brilliant red object. Picking out details on the red planet is never easy, but if you have a medium-sized tele-scope you should see dark markings. These are rocky areas swept clean of dust by wind. On 30–31 January, Mars will be just a tenth of a degree from the Moon, and though the pair will have set in the UK by closest approach, the Americas should have a good view.

On 22 January, Venus and Saturn ap-pear to be close together in the south-western twilight sky (though in reality nearly 1,400 million kilometres/870 million miles apart) and best viewed at around 5pm. This event is called a con-junction and means it will be possible to see both planets together with a pair of binoculars.

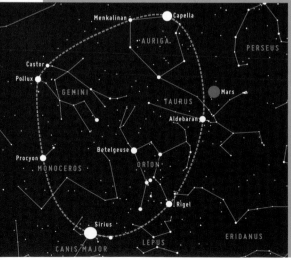

FEBRUARY

It may still be cold in the northern hemisphere, and the winter stars continue to dominate the evening sky, but you don't need to stay up too late to enjoy an astronomical event. And under dark skies you can see the Milky Way running up from the horizon to the left of Orion, the hazy light of billions of stars in our galaxy merged together.

Orion is not only a good target, but a signpost for sights nearby. Betelgeuse lies at the heart of what's known as the Winter Circle, with six more bright stars around it that stand out even in city skies. Rigel is a good starting point, then following the three stars of Orion's belt downwards, you come to Sirius, the brightest star in the sky in the constellation of Canis Major, the Great Dog. Up from Sirius is Procyon, in the Little Dog constellation, Canis Minor. Moving clockwise is Pollux, the brightest star in the constellation of Gemini. Above that is Capella in Auriga, the Charioteer. Finally, further round the circle is the orange star Aldebaran in Taurus, where Mars is still also easy to spot, though now a bit further away.

All of these constellations have objects that are easy to pick out with binoculars or a small telescope. Auriga has the star clusters M36, M37 and M38, the 'M' meaning they are in the catalogue created by Charles Messier in the eighteenth century. All of these, as well as M35 in Gemini, appear through binoculars as glowing hazes of light, with a few stars visible. A small telescope is enough to show tens of stars in each cluster.

PAGE 129: The Milky Way seen over Keswick, in the Lake District. *Stephen Cheatley/Alamy*

OPPOSITE TOP: Orion.

OPPOSITE BOTTOM: Winter Circle.

BELOW: Auriga.

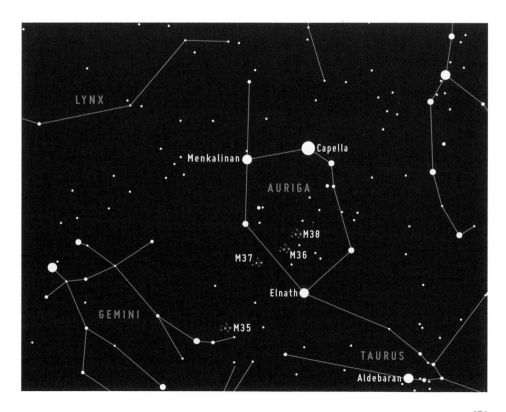

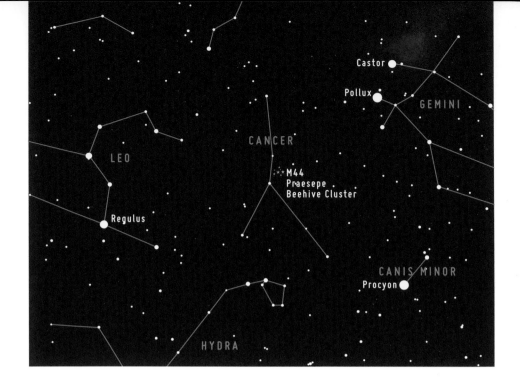

MARCH

By March the days are getting longer and at night the stars of spring are now visible by mid-evening. To the east of Gemini is the faint constellation of Cancer, the crab, with the star cluster M44. This is bright enough to see with the naked eye in a dark sky as an obvious fuzz of light.

Popularly known as the Beehive, M44 was recorded by the ancient Greeks as a 'little cloud', by the Romans as a manger (Praesepe) feeding two donkeys, and by Chinese astronomers as a ghost. The Beehive appears to be about three times the size of the Moon, so binoculars give the best view of the whole cluster. Its 1,000 or so stars (you'll probably see 100 or so of these) are about 600 light-years away.

East from Cancer is Leo, the lion. A reversed question mark of stars forms the lion's head and mane stretching up from the star Regulus, and the triangle to the east marks its crouching hind legs.

The last week of March will be a great time to look at the Moon, when it will be high in the northern hemisphere sky. It starts as a thin crescent just after sunset on 22 March and the visible bright surface broadens over the following days. You should easily be able to see earthshine, where the lunar surface still in night is lit up by light reflected off the Earth. With binoculars or a telescope the mountains and craters stand out along the terminator, the line between night and day.

On 24 March, the Moon will be close to Venus in the sky, and from Asia and Africa will actually appear to move in front of the planet in a rare event known as an occultation.

APRIL

Much of the northern hemisphere is now on summer time, so there's a longer wait for the sky to turn dark, but the usually warmer weather makes being outside more of a pleasure.

This month Venus, the planet closest to the Earth, is easy to spot in the west after sunset. Its thick carbon dioxide atmosphere reflects a lot of sunlight, so it can be as much as twenty-five times brighter than Sirius, and can even be seen during the day. Mercury, the innermost planet, is visible too but harder to see, with the best time being about forty-five minutes after sunset, much lower down in the western sky. Through a telescope both planets show distinct phases like our Moon.

On 23 April the Lyrids meteor shower reaches its peak, with around twelve meteors an hour likely to be visible under dark skies. The shower can produce unusually bright meteors known as fireballs and is best watched before dawn.

Meteors are the result of tiny pieces of interplanetary debris, in this case from the tail of Comet Thatcher that last came near the Earth in 1861, crashing into our atmosphere at high speed. The fragments burn up and usually disintegrate more than 50km (30 miles) above the ground, heating the air around them and creating the light of a meteor. Watching meteor showers is one of the easiest ways to enjoy astronomy, as it's actually better just to use your eyes to see a large area of sky.

OPPOSITE TOP: M44 .

OPPOSITE BOTTOM: A crescent Moon with earthshine visible. *Stephen Rahn*

BELOW: Lyrids.

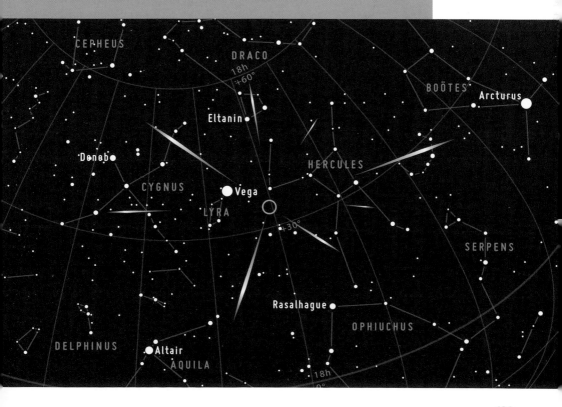

By late evening, in the spring months the stars of Ursa Major, the Great Bear, are almost overhead in the northern hemisphere. The constellation contains the asterism or informal grouping of the Plough or Big Dipper, the seven bright stars with four in a box shape and three in a curved handle. If you follow the curve down you come to the bright star Arcturus in Boötes, the herdsman, then Spica in the zodiac constellation Virgo. The two stars at the western end of the box point up to the Pole Star, which barely moves as the Earth turns.

RIGHT: Ursa Major Pointers.

BELOW: Path of totality for the total solar eclipse of 20 April 2023.

OPPOSITE: A total solar eclipse. *Esteban J. Andrada/CC BY-SA 4.0*

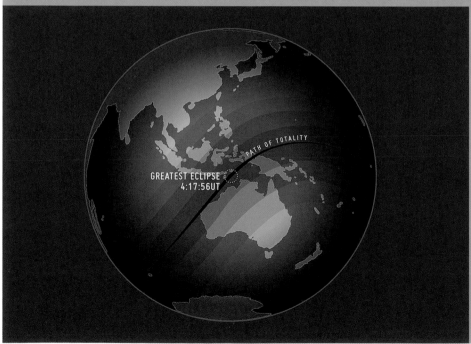

If you're lucky enough to be in the north-western edge of Australia, Kisar Island in Indonesia, East Timor or western New Guinea on 20 April, you'll be in the path of a rare total solar eclipse. The Moon's shadow will move over these locations, where the Sun's bright disk will be completely blocked for about a minute, revealing the beautiful outer atmosphere, or corona.

Except during totality, solar eclipses are dangerous to watch with the naked eye, so either get hold of some certified 'eclipse glasses' with safe filters, project the image of the Sun or, better still, contact a local astronomy society who should be able to help.

"The Sun's bright disk will be completely blocked for about a minute, revealing the beautiful outer atmosphere, or corona."

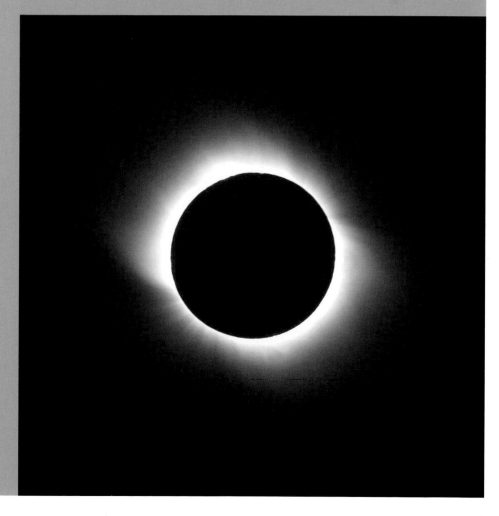

MAY

By the end of May in northern Europe and Canada, the sky is not completely dark even in the middle of the night, as the Sun is never far below the horizon.

Our nearest star, the Sun, is a perfect target in these longer days, and it can be observed using a telescope with a certified solar filter, or by carefully projecting an image on to card using binoculars. The Sun is expected to reach its maximum level of activity sometime between late 2024 and mid-2025, with a larger number of the dark regions called sunspots on its surface. These are about 1,800°C (3,270°F) cooler (but at 3,700°C/6,690°F still hot!) than the material around them and are often as big as the Earth.

In the twilight sky, Venus is very obvious, and at its best at the end of the month. On 20 May, a very thin crescent Moon will be visible, with just 1.4 per cent of the surface we see lit by the Sun. Moons this thin are usually hard to see, and through binoculars or a telescope the crescent is likely to look ragged along the whole of its length.

RIGHT: Solar binocular projection.

OPPOSITE TOP: Hercules star cluster. *ZTF, Giuseppe Donatiello*

OPPOSITE BOTTOM: Summer Triangle.

JUNE

June might not always be warm, but it does have the longest days and the shortest nights. So do take the chance to keep observing the Sun, which is at its highest on the longest day, at the summer solstice on 21 June.

If you do stay up late, Venus is still pretty much at its best this year, is easily the brightest object in the sky after the Sun and Moon, and will be the first 'star' to appear as the sky darkens. As the month progresses, a telescope will show it shrinking to a crescent as it moves between the Earth and the Sun, but its diameter will grow as it gets closer to us.

On 2 and 3 June, Mars will appear to be directly in front of the star cluster M44 (see March), an event that should be visible to the unaided eye and will be a nice sight in a pair of binoculars. The pair will appear low in the western sky as it gets dark.

July sees the nights get a little longer, and it starts to get easier to find the highlights of the summer sky. Three bright stars are a useful pointer to the Milky Way: Altair in Aquila (the eagle), Vega in Lyra (the lyre, an ancient stringed instrument), and Deneb in Cygnus (the swan), together making up the Summer Triangle, a pattern visible through the autumn despite its name.

The band of the Milky Way runs all the way down to the southern horizon to its richest section in the zodiac constellations of Scorpius (the scorpion) and Sagittarius (the archer). This is the direction of the centre of our galaxy, and is packed with stars, nebulae and clusters. It is quite low at northern latitudes, but if you are in the south then the view can be spectacular, even more so with a pair of binoculars.

One standout object is the globular star cluster M13 in Hercules (just to the west

of Lyra), with hundreds of thousands of stars more than 22,000 light-years away. In binoculars it looks like a circular fuzz, and in a medium-sized telescope it resolves into streams of stars, with at least hundreds visible on a good night.

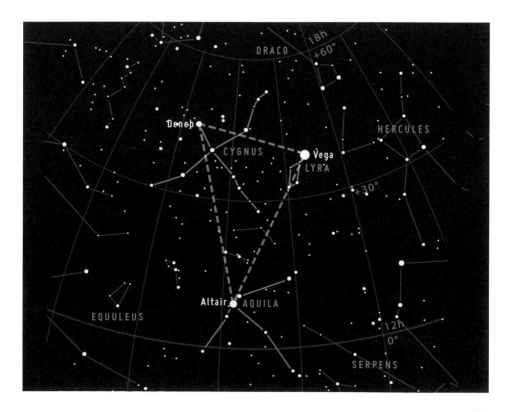

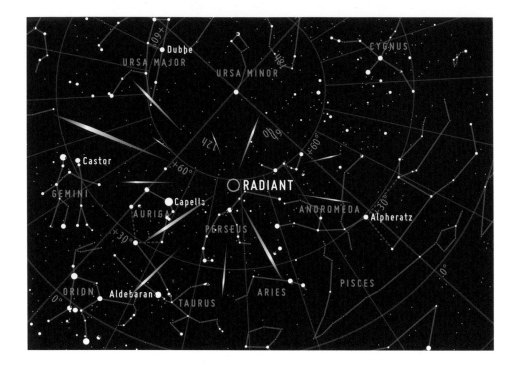

ABOVE: Perseid meteor shower radiant location.

BELOW: An illustration of Saturn.

AUGUST

The Perseid meteor shower is one of the best of the year, and it reaches its peak this month, with maximum activity expected overnight 12–13 August. Debris in the shower is associated with the tail of Comet Swift–Tuttle, which last passed near the Earth in 1992. From mid-northern latitudes the shower is visible for most or all of the night, and meteors should be visible shortly after it gets dark. The best time is likely to be before dawn, and it might be possible to see forty or more meteors an hour, though predicting rates is hard. The good news this time is that the Moon will be a thin crescent, so there will be a real benefit from watching the meteors from a dark sky site.

Saturn comes to opposition on 27 August, meaning it will be opposite the Sun in the sky and highest in the south at midnight (or 1am BST), visible with the naked eye as a bright yellow point in the constellation of Aquarius. The ringed planet will be more than 1.3 billion kilometres (800 million miles) away. Even at that distance it is easy to see the main 270,000km (168,000-mile) wide rings and the 117,000km (73,000-mile) wide planet with a small telescope, always an unforgettable sight.

SEPTEMBER

September can be an ideal month for sky watching, with longer nights and the Milky Way high in the sky. The autumn constellations start to come into view, and the huge Square of Pegasus is prominent later in the evening. The Square is actually split between the constellations of Pegasus and Andromeda, which extends from the top left.

Andromeda has the famous galaxy, M31, just about visible to the eye from dark sky sites, located 2.5 million light-years away, and made up of about a trillion stars. With binoculars it looks like an elongated haze, and with a small telescope its two companions, M32 and M110, come into view.

A harder target this month is the ice giant planet Neptune, which reaches opposition on 19 September, in the constellation of Pisces, underneath Pegasus. On this night it will be 4,300 million kilometres (2,700 million miles) away, and so despite being nearly 50,000 kilometres (31,000 miles) across (four times the diameter of the Earth) it looks like a star in binoculars, and a tiny blue disk in a medium-sized telescope.

BELOW: Andromeda.

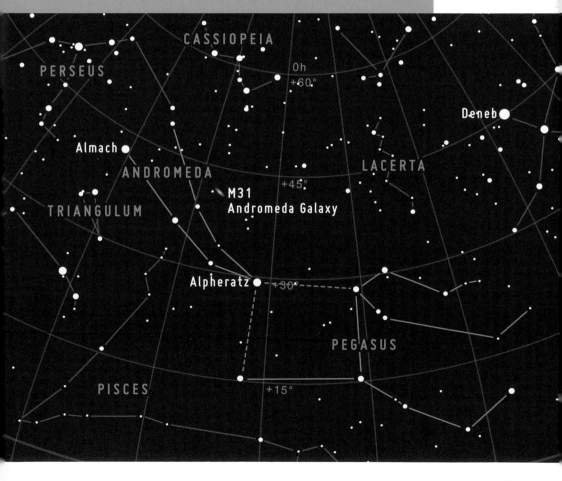

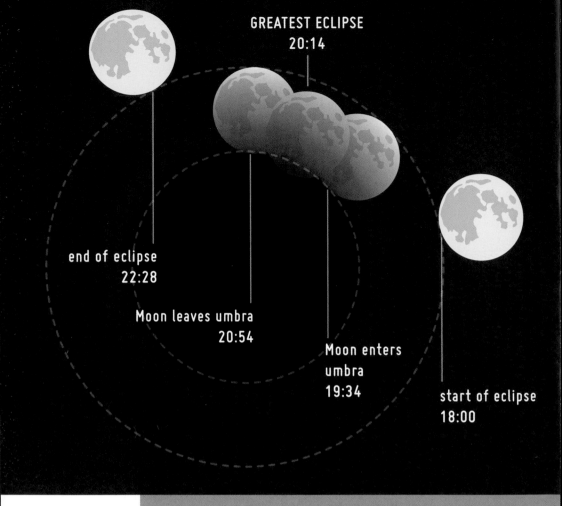

GREATEST ECLIPSE
20:14

end of eclipse
22:28

Moon leaves umbra
20:54

Moon enters
umbra
19:34

start of eclipse
18:00

OCTOBER

This month has both a solar and a lunar eclipse, though mostly visible from different parts of the world. On 14 October the Moon passes directly between the Sun and the Earth. This time the Moon is too far away to completely cover the Sun, with a maximum of 91 per cent of the solar disk blocked out. This leaves a bright ring of sunlight – an annulus – around the lunar silhouette, so the eclipse is annular along a narrow track running from the western United States (including cities like Santa Fe and San Antonio) through Mexico and into Central and South America, as opposed to being a total eclipse. In most of the rest of the two continents, and the western edge of Africa, the Moon will not be lined up with the Sun, so instead a partial solar eclipse will be visible, where a bite will appear to be missing from the solar disk. If you plan to watch this, do please follow the safety advice given for the April eclipse.

ABOVE: Timings for the partial lunar eclipse of 28 October 2023 visible from Europe, Asia and Africa. Timings are in GMT.

OPPOSITE TOP: The Double Cluster in Perseus. *Stephen Rahn*

OPPOSITE BOTTOM: Perseus and Cassiopeia.

Two weeks later, on 28 October, Europe, including the UK and Ireland, most of Asia and most of Africa will see a partial lunar eclipse. In western Europe, the Moon will be in the evening sky when just 6 per cent of the sunlit surface enters the darkest part of the shadow (the 'umbral shadow') of the Earth. Most of the Moon will have a yellowish hue from the outer and lighter terrestrial shadow, the penumbra, and a small nick of darker shadow will be visible.

Throughout the month, look out for constellations like Perseus, rising in the east as the Sun sets, and the distinctive 'W' shape of Cassiopeia above it. At the top of Perseus is the Double Cluster, NGC 869 and NGC 884. These two star clusters are about 7,500 light-years away. The clusters are just bright enough to be seen with the naked eye, and with binoculars or a small telescope are one of the most spectacular sights in the sky, with hundreds of stars visible.

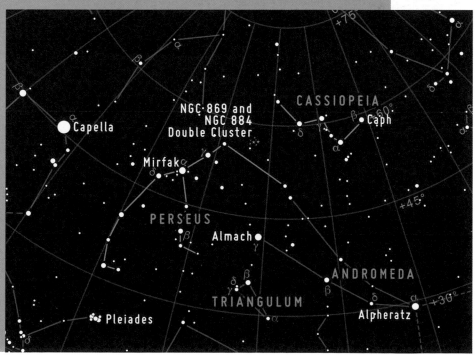

NOVEMBER

November is the best month this year to look at Jupiter, the largest planet in the Solar System. The gas giant is at opposition on 3 November, when it will be slightly less than 600 million kilometres (370 million miles) away, and brighter than any star. By midnight Jupiter will be high in Aries, the ram, a small constellation to the west of Taurus.

Much fainter and harder to see is the ice giant Uranus, similar in size to Neptune, which reaches opposition on 13 November, also in Aries. The distant world will be 2,800 million kilometres (1,740 million miles) away, appearing starlike in binoculars and as a tiny green-blue disk with a good telescope.

By November, Taurus is easy to see again. With the exception of the bright star Aldebaran, the stars of the 'V' that mark the horns of the bull are actually members of one of the closest star clusters, the Hyades. Higher up towards the west is another cluster, the Pleiades or Seven Sisters, represented by the seven stars visible to the naked eye. Both of these objects are impressive in binoculars, which reveal many more stars.

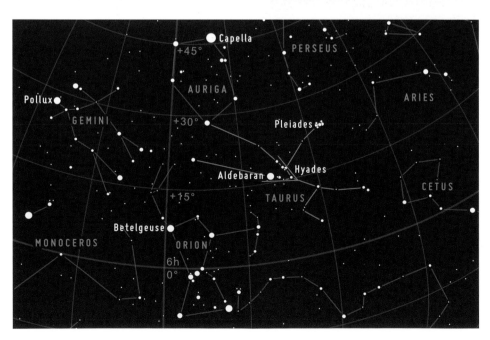

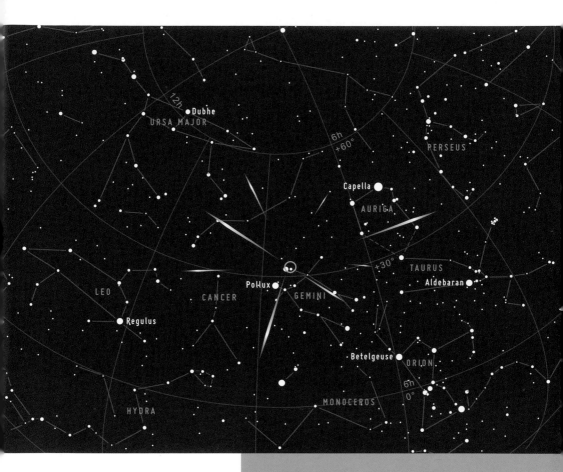

DECEMBER

As the temperature drops in the north and we embrace the cold, clear nights, the bright winter stars are again the main feature of the sky, with Orion and the Winter Circle dominating later in the evening.

On the night of 14–15 December, there is an additional treat as the Geminids meteor shower reaches maximum activity. The Moon will be new, making it worth the effort to travel to dark skies to watch. This shower is typically the strongest of the year, and at its peak you could see more than sixty meteors an hour.

"At its peak you could see more than sixty meteors an hour."

OPPOSITE TOP: An illustration of Jupiter.
OPPOSITE MIDDLE: Pleiades. *ESO/S. Brunier*
OPPOSITE BOTTOM: Taurus.
ABOVE: Geminids.

GETTING STARTED
WITH A TELESCOPE

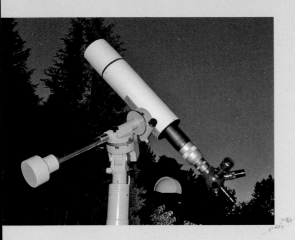

You don't need an expensive telescope to look at the stars and planets. Astronomer Robert Massey has put together this short guide to help you get started.

BINOCULARS

Binoculars are by far the best way to start out in astronomy. They are affordable, portable, easy to use and versatile. With even a modest-sized pair you can see craters and mountains on the Moon, the star fields of the Milky Way, the moons of Jupiter, the Orion Nebula and the Andromeda galaxy.

Most pairs of binoculars will have the magnification and the diameter of the main (objective) lenses written on them in millimetres, usually in the form 'number x number'. Dividing the objective by the magnification gives the 'exit pupil' in millimetres, a measure of how wide the pencil of light is that reaches our eyes. As a rule, about 5mm is right (from 10x50 or 8x40 binoculars, for example), as that matches the size of the pupil of the eye in typical dark conditions. But if your binoculars have a smaller pupil, don't worry, you can still see good views of many sky objects.

Try out binoculars during the day by looking at a distant object (although remember never to look at the Sun). You should be able to focus with the centre wheel, and the right eyepiece can then normally be adjusted to compensate for the difference between your eyes. At night it can be slightly more difficult to point at stars, so one trick I use is to look at the object I'm interested in and then move the binoculars in front of my eyes, a bit like putting on a pair of spectacles.

With binoculars it really helps to have a tripod too, particularly if they're a little heavy, or if you're trying to show the view to other people. Using a tripod will give you stability and also has the advantage of being 'hands-free', allowing you to refer to any books – like this one – or sky charts that you may have with you. Most pairs have a simple screw hole where an adaptor can be attached.

TELESCOPES

Telescopes are a bit harder to set up, but can be much more powerful. They are either refractors, which use a lens to gather light, or reflectors, which use a concave mirror. All telescopes for personal use then have a second lens where the light is directed into your eye, which is called the eyepiece. Telescopes designed for astronomy often give an inverted view, so astronomical targets appear upside down. Correcting this requires additional lenses.

While the most advanced telescopes for amateur astronomy cost tens of thousands of pounds, those bought by beginners or as gifts are usually much cheaper, yet should still be able to show you sights like Saturn's rings or the dark bands on Jupiter.

A telescope must be mounted on a tripod as they usually have a much higher magnification than binoculars, and are generally larger and heavier, so are very difficult to hold steady. One of the first things to check when you have a new telescope is the tripod, so try moving the telescope around, and make sure that the clamps are tightened enough so it doesn't wobble. The simplest tripods are like those used with cameras. They move up and down (altitude) and left to right (azimuth). These are called altazimuth mounts.

These are fine for many purposes (I have two telescopes with these mounts) but they make it slightly harder to follow stars and planets across the sky as the Earth turns. Other telescopes have equatorial mounts, where one axis of rotation can be aligned with the axis of the Earth, and the telescope can then follow stars with a single movement.

Most telescopes will have what looks like a smaller telescope mounted on the side. These are finderscopes that have a wide field of view and low magnification with crosshairs to help you find an object before looking at it through the main instrument. The main scope and finderscope need to be lined up, which is a lot easier to do during the day. Point the main telescope at a distant object and bring it to a focus. Patiently adjust the small screws on the finderscope until the same object appears in the centre of the crosshairs, and retighten the screws. Provided the finderscope stays in place, you can then use it to find targets at night. More sophisticated telescopes have electronic guidance systems, but setting these up needs a basic knowledge of the sky and where the brighter stars are found.

Another key property of a telescope and its eyepieces is the focal length. This is essentially the distance for which light from a distant object (like a star or planet) is brought to a focus, and is usually given in millimetres. The magnification is then the focal length of the telescope divided by the focal length of the eyepiece. A 1,000mm telescope with a 10mm eyepiece would give 100x magnification, for example.

It always helps to start with a low magnification eyepiece to find an object and then to increase the magnification to see more detail. A rough guide is that the maximum usable magnification is about twice the size of the lens or mirror in millimetres, so a 60mm refractor will not work well above 120x. Planets usually need a higher magnification, whereas objects like clusters or stars or nebulae need a wider view.

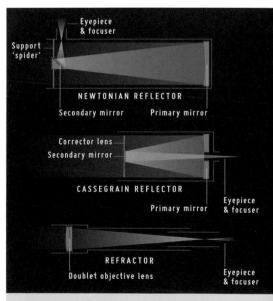

ABOVE: The different configurations of reflector and refractor telescope.
OPPOSITE: A typical refractor telescope. *Sternwarte-Weinheim/Alamy*

TO THE

Climate change, war, famine, inequality – is life on Earth getting you down? Why not apply to join the Supermassive Space Ark™ voyage to Earth 2.0, where humans will be able to start again? Help us build a new, improved and equitable civilisation. Richard Hollingham offers a vision of the future.

STARS!

T he sky switches on at 6am. The bright, uninterrupted blue stretches to the horizon, dotted here and there with perfect light fluffy clouds. It's going to be another glorious day.

Just behind the rows of corn, swaying slightly in the gentle breeze, two figures in ragged overalls trudge across the field. Beyond them, in a small wooden shack, the children are getting dressed. No school for them; they will be helping their parents with the harvest.

The bees come online at 6.30am, pollinating the fruit trees in the next field. A few don't make it. One day they will fail completely.

It's fifty years since the Supermassive Space Ark™ left Earth's orbit to travel beyond the Solar System on a mission to colonise a distant world around an alien star. When they signed up, the hundreds of carefully screened volunteers on board this interstellar spaceship accepted that it could be several genera-tions before they arrived at their new home. Along the way they would need

to navigate and maintain the spacecraft, grow food, recycle water and monitor the air. It would be their grandchildren who would take the first steps on Earth 2.0.

But what began as a democratic society when the Supermassive set off – in which even the corporation's CEO mucked in with the harvest – has descended into feudalism. The CEO's daughter is now captain, and the children of the farmers on deck nine are destined to spend their life in the fields under an artificial sky. The engineering know-how to fix the bees has been lost.

'The culture of an interstellar spaceship has the potential to become a police state,' warns Les Johnson, a physicist and author, with a day job as a senior NASA engineer. Currently working on a forthcoming robotic asteroid mission, Johnson is a leading member of the US-based Interstellar Research Group. The think-tank advocates for interstellar exploration and is developing the plans to help make it possible.

But despite being in favour of humans leaving Earth permanently, Johnson is all too aware of the downsides of locking up a bunch of people in a giant spaceship. 'You don't want to set up an inherited feudal system but, at the same time, you've got to have someone to fix the toilets.'

COSMIC CATASTROPHE

The idea of taking human civilisation to the stars, to set up life on distant worlds (and perhaps make a better job of it than we have here) is a beguiling one. And, to many of its proponents, even an essential one. The first reason you might want to leave Earth is survival. You can buy a T-shirt with the slogan 'The dinosaurs didn't have a space programme', the implication being that only space can save us from eventual annihilation.

If we don't manage to destroy ourselves through messing with the climate or a devastating nuclear conflict then, the argument goes, we could be obliterated by an asteroid or other unexpected cosmic catastrophe. Ultimately – albeit in around 5 billion years' time – the Sun will run out of hydrogen to burn and become a red giant, subsuming the Earth. Sending a bunch of humans deep into space is like launching a life raft from a sinking ship – at least there will be some survivors.

Another reason to bravely go where no one has gone before is because exploring new frontiers is what humans do. If it wasn't for our yearning to explore, the first tribes of *Homo sapiens* would never have left Africa to spread around the world and ultimately shape the Earth to our needs. We certainly would not have sent rovers to Mars or probes to investigate the depths of the Solar System. The oceans were once the final frontier – now it's space.

There is also the chance that, even if life is common in the Universe, intelligent life forms are rare. We could be the brightest beings in the Universe, so perhaps reaching for the stars is our destiny.

'We have a moral obligation as humans to spread ourselves, spread life, spread our culture, the best of us,' argues Johnson. 'If the Covid-19 pandemic has taught us anything, it's that existence can be tenuous.'

So, if that's the case, how do we go about building the Supermassive Space Ark™, where will it go, and what guarantees can we give to the volunteers who sign on?

"Sending a bunch of humans deep into space is like launching a life raft from a sinking ship."

PAGE 146–147: A future Martian habitat could house hundreds of people. *Artistic rendering: Dotted Zebra/Alamy*

OPPOSITE: An artist's impression of an exoplanet seen from an alien moon. *IAU/L. Calçada*

BELOW: A more industrial-style concept for a Mars base. *Artistic rendering: 3000ad/Alamy*

VOYAGER

The good news is that humans have already achieved interstellar travel. In 1977, NASA launched the twin *Voyager* probes on a journey to explore the planets beyond Mars. After flying past Jupiter in 1979 and Saturn in 1980, beaming back detailed images of the gas giants and their moons, *Voyager 1* kept on going until, in 2012, it left the Solar System entirely. Travelling at some 61,000km/h (38,000mph), it is now almost 24 billion kilometres (15 billion miles) from Earth.

Voyager 2, meanwhile, launched on a different trajectory towards Jupiter before flying past – for the first and still only time – Uranus and Neptune. It left the Solar System in 2018. Although many of their instruments have been shut down, after forty-five years in space, the twin probes are still sending back data drawing just a few watts from their nuclear batteries.

Voyager 1 and *Voyager 2* are more than scientific explorers; they are also our ambassadors to the stars. Each spacecraft carries engraved pictures of humans, the position of Earth in the galaxy and a golden record containing images and sounds of our world. The soundtrack covers everything from birds to Bach, and rain to rock and roll.

Unfortunately, however, given their speed, trajectory and the fact that there is a great deal of nothingness in space, neither *Voyager* will reach another star system for at least half a million years. It's unlikely any alien beings are going to be listening to 'Johnny B. Goode' by Chuck Berry anytime soon. And, for those signing up for the Supermassive Space Ark™, half a million years is not an ideal timescale for any journey. (To put it into perspective, humans were in the Stone Age half a million years ago; the first metalwork with bronze did not start until around 3300 BC.) So, what are our options for getting to a new world a little bit quicker?

EARTH 2.0

The first challenge for our intrepid inter-planetary explorers will be deciding on a destination. When the *Voyager* probes set off, astronomers could observe plenty of stars but had yet to prove there were any planets beyond our Solar System. Since the launch of NASA's Kepler space telescope in 2009, our knowledge of exoplanets – planets that orbit stars outside our Solar System – has been transformed. At the time of writing, we know of more than 5,000 confirmed exoplanets, with at least another 8,000 potential candidates.

Many of these newly discovered planets are gas giants like Jupiter, but some are almost certainly rocky like Earth. With JWST (see page 8) now beginning its observations, we can expect many more exoplanet revelations. The new telescope will also be able to help discern the atmo-spheres of planets and whether they might be suitable for life.

It's not therefore unreasonable to as-sume that within the next few decades we can find an Earth 2.0 – a planet that is suit-able to be seeded with life or, ideally, one already rich with vegetation. The biggest technical challenge is going to be develop-ing a spacecraft that can get there within a reasonable timeframe – a few generations, say – without killing the crew or running out of fuel during the journey, leaving it to drift aimlessly in the void until the ship finally falls apart.

According to Albert Einstein, nothing in the Universe can travel faster than the speed of light. Light travels at 299,792 kilometres per second (186,282 miles per second), which is exceptionally fast, but, as Einstein pointed out, everything is relative.

'If the speed limit is the speed of light, then just getting to the nearest star is over a four-year journey,' says Johnson. 'The chances are, we won't be going at the speed of light, and chances are there isn't a hab-itable world that's that close. This means they're going to be farther away, which could be up to 100 light-years.'

ENGAGE WARP DRIVE

Options for propelling an interstellar space-craft include using unproven – but not impossible – technologies such as nuclear fusion generators where atoms are fused together to create energy, the same process that powers the Sun. The designers might also deploy solar sails – giant shiny wings that are pushed along by photons from the Sun or even propelled by lasers fired from Earth.

But hang on, you say ... what about warp drives and wormholes, the staples of space travel in almost all science fiction? *Star Trek* would be exceptionally dull if it took the *Enterprise* 100 years to travel between strange new worlds.

Einstein described how space and time are fused together and can be warped by mass and energy. Planets, stars and black

holes curve space-time around them, which we experience as gravity. The idea behind a warp drive is to compress space in front of your starship, effectively reducing the distance you need to travel so you can move across space in less time. The theory is sound, and several physicists have done the maths. But calculations suggest that to generate the required energy to warp space you would need a mass the size of the Sun, so the practicalities need some work.

The theory of wormholes was conceived by Einstein and Nathan Rosen as tunnels across the Universe, offering shortcuts between the stars. Again, in theory at least, wormholes could exist in the fabric of space-time, but the jury's still out on whether they are feasible or practical to use. Which leaves us, at any rate for now, with plan A – a ship that goes as fast as possible but still takes

decades, maybe a hundred years, to reach its destination. So here's another suggestion: why not sleep through it all?

THE BIG SLEEP

Any half-decent science-fictional starship that's not been fitted with warp or worm drive usually employs suspended animation cocoons instead, allowing the crew to sleep through the voyage and arrive refreshed at their destination.

'The crew are always beautiful and they're in their twenties and they wake up and they're still in their twenties, and the guy just has a little bit of stubbly growth on his beard,' says Johnson. 'I think that's pretty fanciful – even if we were able to find a way to suspend animation, to slow down metabolism, you'd probably still age and that means

you couldn't be in a box for a hundred years because you'd grow old and die.'

Cheery stuff. You might perhaps expect more optimism from an advocate of interstellar travel. But that's the point. If the Supermassive is to succeed, then as far as Johnson's concerned, any design has to be rooted in the real science of the Universe we live in, not the imagination of Hollywood screenwriters.

'You could probably help relieve the monotony of the voyage by finding a way for people to be kept in a relatively suspended animation where the metabolism slows down,' Johnson says. 'They aren't consuming as many resources on the trip, like air, water and food. But I think that the science-fictional notion that you can freeze people and they stop ageing ... I've not talked to anyone that really believes that's possible.'

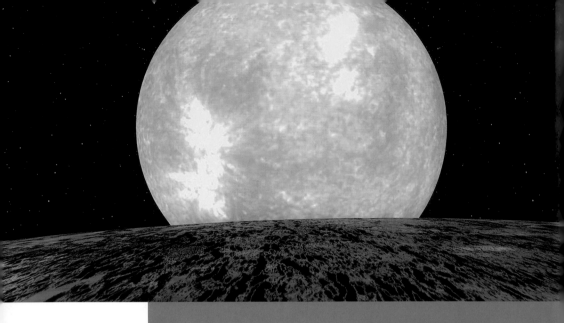

LIVING STARSHIP

So, let's get into the small print of the terms and conditions for the Supermassive Space Ark™. It's going to be a long voyage over generations in a ship, with a society that may descend into a police state, to a destination that we have yet to discover. You might be able to sleep for part of the journey, but you will still age and it will be your children or grandchildren who get to make a new life on a new world, not you. Oh, and you can forget the idea of living in a shiny starship like the *Enterprise* with food replicators, holodecks and gleaming white corridors.

'The amount of food that we would need means we would need to produce our own food,' says Rachel Armstrong, a professor of regenerative architecture at the prestigious KU Leuven University in Belgium. 'We would need an environment that actually sustains us and isn't just a temporary holding space.'

A leading thinker in the area of interstellar space travel, Armstrong has helped develop the idea of a living starship, where biology is integrated into the entire structure to create an artificial biosphere. She imagines using the ship's fusion reactor to heat hot springs to culture microbial soups, soils rich with bacteria and a network of fungal mycelia. There will be reed beds to recycle human waste back to drinking water and mats of vegetation to clean the air.

'Unlike the images that we get of space travel, I think we will be more agrarian, tending to the land in our daily lives,' she says. 'We may well be wading through mud and warm water, looking to see which crops are taking, looking to see which cell cultures are showing signs of stress and trying to figure out how we might do things differently.'

PAGES 152–153: Warp drives have captured the imagination of countless science-fiction writers. *Artistic rendering: Vitaly Sosnovskiy/Alamy*

ABOVE: Artist's impression of the view from planet Kepler 10-B. *NASA/Kepler Mission/Dana Berry*

OPPOSITE: Artist's impression of a transiting exoplanet. *ESA/C. Carreau*

"It's going to be a long voyage over generations ...
to a destination we have yet to discover."

GLOOP

Recreating the complex biological systems we all rely on is going to be challenging. Just growing a plant in a greenhouse or garden is hard enough and previous attempts to build large-scale biospheres on Earth have had mixed results. The most famous, a 1990s experiment called Biosphere 2, was built as a self-sustaining complex in the Arizona desert. After a year, crops were failing, and the oxygen levels had dipped to dangerously low levels. Armstrong suggests that those on board any future starship will need to be connected to the biological systems around them in ways that most of us have lost touch with on Earth.

'I think we will get quite neurotic if we start to see signs of stress in some of the microbial cultures and I think that we will see our futures in them in some ways,' she says. 'I think we're going to be emotionally involved with these collections more than we are on planet Earth because there's a huge amount at stake.'

As for food, you can probably guess where this is heading. Although some food might be grown in fields, basic sustenance will likely be based on microbial and algal soups. It's unlikely there will be much, if any, meat on board, although there might be fish. So, unless you are sitting at the captain's table, the dish of the day is probably going to be more gloop than gourmet.

So, are you now tempted to put in your application and join the rigorous selection process for the Supermassive Space Ark™? Sadly, we don't yet have the funds to make it a reality. But, despite all the challenges, I can guarantee that, if the project really existed, hundreds of thousands of people would put their names forward. A recent – failed – project called Mars One, for example, aimed at flying volunteers on a one-way trip to Mars to establish the first Martian colony, attracted more than 10,000 potential astronauts.

If you do sign on for the ultimate human adventure, just remember that terms and conditions do apply.

lava life

Skies sparkle above a never-ending ocean of lava

"Are you ready to sign on for the ultimate human adventure?"

OPPOSITE: Poster from NASA's 'Exoplanet Travel Bureau', extolling the virtues of Kepler-16b with its double sunset. *Artistic rendering: NASA-JPL/Caltech*

RIGHT: Planet Janssen, or 55 Cancri e, is only 41 light-years away. Fancy a holiday riding above its molten surface? *Artistic rendering: NASA-JPL/Caltech*

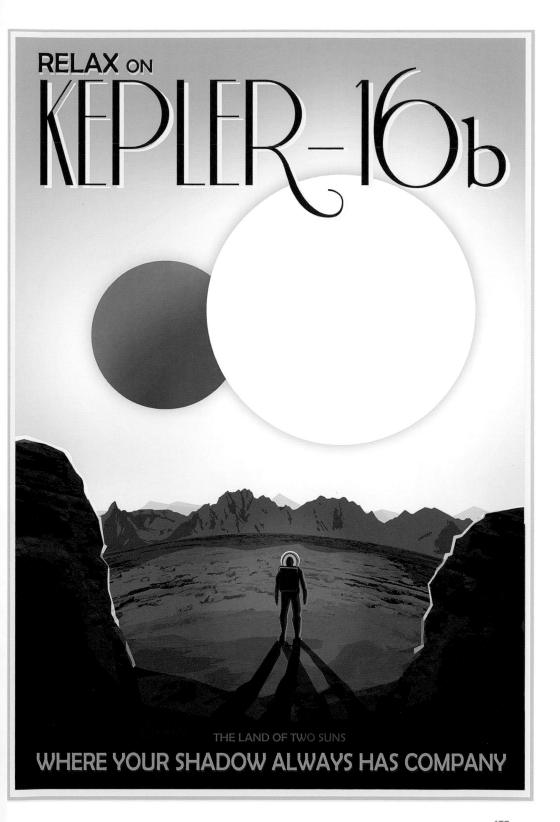

RELAX ON

KEPLER-16b

THE LAND OF TWO SUNS

WHERE YOUR SHADOW ALWAYS HAS COMPANY

INDEX

Page numbers in **bold** refer to illustrations and their captions.